CW00487152

Of Earth, For Earth.

The meaning of a mine.

Of Earth, For Earth
The meaning of a mine

Edited by Kathryn Moore, Dana Finch and Bridget Storrie

Book design and additional images by Owen Tozer

This edition published 2020
by the IMP@CT Project

ISBN: 978-1-5272-7662-8

Copyright: © IMP@CT Project
All rights reserved

Printed and bound by:
Short Run Press Ltd
25 Bittern Road
Exeter EX2 7LW

In memory of Dylan McFarlane

Contents

FOREWORD

Of Earth, For Earth began as an art exhibition to inspire debate about human interactions with the Earth and materials of the Earth. The IMP@CT project published an open call to engage professional artists in a dialogue about mining and the Earth more generally. The response was enthusiastic and gratifying, and it engaged with the suggested themes: mining for raw materials; telluric voids; the meaning of a mine; narratives of extraction; the conversation between self, material consumption and Earth. Six exhibiting artists took the themes and created something uniquely questioning. The old engine house at the Heartlands historic mine site was the venue for the exhibition, complete with train tracks and bits of rusty iron lifting equipment hanging from the ceiling, such that the art space colluded to create a narrative beyond the sum of its parts. The artists conferenced with scientists, social scientists, engineers and miners. The conversations about mining took as many twists and turns as an underground seam, many inspired by the narrative created in the exhibition. We were led into ever deepening mental and visual labyrinths, discussing the poetics of underground space, the meaning of a mine from a variety of perspectives, the social cohesiveness possible in underground spaces that might vanish on the surface, the elements of danger, trust and identity. With the cooperation of additional artists and writers the conversations coalesced into this book. In many ways the meaning of the word mine encapsulates all.

The pronoun mine is defined as 'belonging to me'. In this book it speaks to our sense of belonging to place, time, natural and cultural heritage - if these can indeed be separated. It speaks to our attachment to our planet, our environ, our homeland, the further landscapes we choose to access or seek to protect. It speaks to the way in which we try and assert our ownership over Earth: purchasing plots of land to call our own; using it for the sustenance on which our very lives depend; exploiting Earth materials. It speaks to the resources we use to add value or comfort to our existence, and that enable us to express ourselves through the medium of art. It speaks to our consumption of Earth, our alteration of Earth, our appreciation of Earth. The section 'Of Earth' includes voices that speak of connection, individual or collective concern, recognition of our own failings in our relationship with Earth resources and commitment to change.

The noun mine is defined as a 'deep hole for digging out coal, metals, etc.' or a 'hidden deposit of explosive'. It is a fixed thing, geologically grounded. The definitions initially bring to mind two different scenarios: the first of industry and commerce, the second of conflict. But both definitions are situated in place and both infer the removal and/or destruction of Earth. Ore deposits are situated by the geological forces behind their creation. They cannot be moved out of the path of conflict. The right to mine ores may be coveted by adversaries, and the industrial output can be used to finance conflict, such that industrial mining operations can become contested places. But the mine does not need to be a negative space – it can be more than a telluric void whether occupied or abandoned. It can be a place of regeneration, of healing. The section 'For Earth' articulates connection to place, the positive expectations that arise from the promise of a mine, the negative experiences that can be associated with proximity to a mine, hope for what can be achieved in the regeneration or afterlife of a mine.

The verb mine is defined as to 'dig from a mine' or to 'place explosive mines in'. It encompasses the processes of removal, of extraction, of reshaping, of destruction. By utilisation of explosive technology, mining spectacularly expanded in early modern history. Mining is entwined with innovation in wider industries: the use of exploration techniques developed for military purposes, or digital

'Under the mountain', oil on panel, 2020 Dana Finch.

techniques for automation in recent mega-mines. Mining is an ancient activity that adapts and evolves, permeating society with the materials for industrial and economic development. Mining activity shapes our landscapes. It imbues those working in the mining sector with a sense of belonging to an industrial community with thousands of years of heritage. The echoes of time are articulated by many of the contributors to this book. We return to the locales of mining. We learn from the impacts of both past mining and poor practice in recent/active mining. We try to evolve, to apply better the methodologies that we currently have at our disposal, and to find new methodologies and ways of operating. The section 'Un Earth' describes experiences, learnings and attitudes towards the action of removing material from the Earth. More than that, it describes hope that we may unearth mechanisms to carve out a more sustainable relationship with the Earth while maintaining access to the resources that will support the global population.

Kathryn Moore, IMP@CT project lead.
Dana Finch, IMP@CT project manager.
Bridget Storrie, consultant mediator & researcher.

Installation views, Of Earth, For Earth exhibition, 2020.

H E A R T

This book is full of images from the Of Earth, For Earth exhibition. A group exhibition is not just a collection of individual art works; it is a story, with no beginning or end, but an ongoing push and pull between the characters, and multiple tensions and harmonies rising and falling as the viewer wanders through it.

So, in a non-linear way, this story began with James Hankey's dust >| dust, a film piece, shown in a dark space next to the main gallery room. The seven minute film played on a loop, and took us into a contemplative state, as rocks of varying sizes and degrees of brightness cascaded into a black void. It was mesmerizing and strange, accompanied at times by crashing sounds as the rocks hit the sides or bottom of whatever they were falling into, and at times utterly silent, leaving the viewer to wonder if there was any end to the void. In the slow and constant movement, time stood still. Only shape and darkness, light and absence of light, existed. Something was connecting the surface to the inside of the Earth: our trail of perception, captured in the falling objects. Rocks and stones, white, pale, elongated, rounded, jagged, tiny and massive, tumbled before our eyes. An occasional twig joined the flood of matter, spiralling into darkness, into this telluric void.

It reminds us that Cornwall is full of holes, left by centuries of mine workings. Wandering around on the hard, granite surface, it is hard to imagine that beneath our feet there is a lattice of tunnels at varying depths and angles, leaving us suspended precariously over this labyrinthine honeycomb of fragility and absence. It seems miraculous that the whole county doesn't collapse in on itself, so comprehensively has the ground been hollowed out.

A falling into somewhere introduces the story, and begins our journey. Next, we come upon a kaleidoscope, in fact the whole room is a kaleidoscope, glimpses of fractured light suspended in a crepuscular space. We have to make sense of this, and it takes some scrutiny. An actual kaleidoscope, created by Josie Purcell, stands in its own cabinet, but you are invited to take it out, hold it to the light and make your own crystallised world from its dancing molecules. Its metal casing is both baroque and luxurious, speaking of the riches of gemstones and precious metals. Purcell's series of jewelled images, 'Elemental Eidos' draw the viewer further into the story; we are going underground, deeper into the mechanisms of the Earth at a molecular level. The falling rocks have solidified into crystal, kaleidoscopic images, in intense deep colours. They vibrate across the space, each in perfect symmetry, made from sand, and minerals and a little bit of light, using the processes of cathodoluminescence and lumen imaging.

In the centre of the wall hangs an enormous oval, sundered down the centre to reveal the organic lattice of life. Dan Pyne's 'Costing the Earth' invites us to consider both the material and the moral issues involved in creating art. It is an imposing image, an Earth ovoid split in two, spilling its innards and yet maintaining its visual integrity through tensile harmonies, while acknowledging the fracture at the heart of our relationship with the planet. On the adjacent wall is a print of the same image, made using Bideford Black, a pigment Dan took himself from the local ground. Dan completes this work with an impressive list of materials, their source and their rough cost in monetary terms. It is a piece of art which invites real interrogation of the way we use the 'common' materials of Earth.

L A N D S

Occupying the space in the (almost) centre of the room is Jack Hiron's installation 'Utopia, it's down here somewhere', an elegant and humorous assemblage of miners' yellow helmets, lamps and stickers on tall metal poles, and lit mainly from their own helmet lamps, each one illuminating the next, spilling into the surrounding darkness and creating unexpected shadows throughout the exhibition. Our story has become human. Mining throughout the millennia has created culture. Mining stickers became a way to identify certain pits, and each has a meaning. Hirons has taken them and subverted the messages, to confront us with easily dismissed statements such as 'Don't quarry, be happy', that is, until the realisation of multiple meanings sinks in.

Remaining in the human dimension, we encounter the work of Henrietta Simson whose painting of a medieval Renaissance landscape, entitled 'After Thebaid' echoes the clay pit terrain of mid-Cornwall. A bright silver river flows through the work, slicing the canvas, catching our attention and drawing us in. Around this large painting are others, smaller, punctuating the wall with cyphers of landscape; a cave, a cloud, a rock, each with many narrative interpretations, all deliberately refusing to conform to the conventions of western aesthetics. On the raised floor are two plinths, for the works 'Material Culture 1 and 'Material Culture 2' containing a selection of ceramic forms, sculpted into irregular geometries, echoes of the Renaissance view of the hills and valleys that provided the space inhabited by the human, often religious, actors in their paintings. Simson's use of materials reflects the content of her work, creating a powerful symbiotic relationship between form and content, especially prominent in the blue and white 'tiled' mountain form which alludes to the

glazes used in tile making across the globe. The glazes, shapes, even the surface on which they are displayed, are all lustrous and tactile, and refer to the earth from which they are drawn.

Finally, we return to the ground itself, to its form and essence, in the three-dimensional textile work of Heidi Flaxman, and her sculptural reimagining of Carn Brea, a hill actually visible from the exhibition site, rising above the town of Redruth. Heidi has created an exquisite set of tapestries, each embodying the geological and topographical constitution of sections of the hill. The intense colours and detailed woven surfaces engage us in an examination of the wonders of geology and structure. Threads run from the work, not unravelling but leading, inviting, and reminding us that we are part of the fabric of the whole. The structures on which the tapestry sits are dark and mysterious, bringing our attention back to the voids beneath our feet.

Together the works are bound into a story; they are held together by their own narrative, and by the space that holds them in a web of shadows and light and the echoes of past mine workings and industrial structures, and they open a new way in to understanding the underground and all its present day connotations.

Dana Finch, IMP@CT project manager.

Section 1.

Of
Earth.

Kathryn Moore.
Of Earth, For Earth.

'Lode', felt, 2020 Kathryn Moore.

'Chemical' dyes have a vibrancy and durability beyond that of natural (plant and animal) dyes, such that they have been attractive to artists for centuries. Many chemical dyes derive from mineral ores that occur in lodes: veins of metal ore in the earth. The colours that are depicted derive from the natural reaction of water with iron- and copper-bearing minerals in a polymetallic lode that is exposed in an abandoned mine face (Outukumpu Mining Museum). Metallic fibres were added to the dyed wool fibre prior to the felting process.

Modern society is riddled with disconnection and paradox. Disconnection embodied in the desire of consumers to acquire finished, packaged goods for a comfortable standard of living, distanced from the realities of raw materials production. Paradox in the perception of living working landscapes as rural idylls that embody nature, even where they are sculpted by human hands. Disconnection and paradox can be dangerous where they breed misunderstanding and mistrust, create barriers to communication, or create feelings of helplessness in the face of (climate) change. I don't believe this is going too far. In my professional life, I strive to find ways to ensure supplies of the raw materials used to build the infrastructure for the transition to sustainable, low-carbon energy futures. I meet individuals committed to limiting global warming, who perceive that mining must be stopped. Yet we cannot achieve the green energy transition without mining. A new dialogue is needed. A dialogue that enables all voices and all concerns to be heard equally. This is a narrative of individuals, whose opinions are formed through their morals and life experience. Upbringing, education, community, livelihood, life-style and personal experience all shape our relationship with the Earth. We use raw materials of the Earth and create infrastructure for the technologies that can limit our impact on the Earth. We are 'of Earth' and 'for Earth'.

I was asked at a public forum *How do we solve the problem of mining?'* I replied with another question *'Is mining a problem?'* Confused faces looked at me blankly, not knowing how to respond. Of course some mining is a problem. When mining goes wrong, it can go spectacularly wrong. It can result in significant loss of life, create suffering on a scale akin to that of a natural disaster, and impact landscapes beyond natural recovery. Environmental scars exemplify the perceptions of some that extractivism inflicts bodily harm upon the Earth. I simply ask the question whether this devastating image is a true reflection of all mining operations. It is important to remember that miners inhabit their working landscapes. They are connected to one another in community and to their environment. I do not romanticise the relationship between miners and environment. In many respects, custodianship of the landscape is limited by economic and political paradigms. The protection of environment, miners and local communities comes at a cost that is in tension with the desire of end-use consumers to obtain the best value for money. Capitalism demands that industry must operate at a profit at every stage of the value chain from the extraction of raw materials to distribution of manufactured goods. If we recognise that the consumption patterns of society are the ultimate control on extraction of Earth materials, there is no *'them'* responsible for mining. There is only *'us'*. We are 'of Earth and 'for Earth'.

We are all affected by mining, directly or indirectly. We are all stakeholders and we have a voice. We use that voice in different ways, in deciding on our consumption patterns, in deciding on our belief systems, in our voting and campaigning. We voice our concerns. But are our voices in conflict or conversation? Do we acknowledge that society is underpinned by materiality? Do we reminisce or regret past industrial endeavour? Do we rail against or applaud the capitalist, consumption-based present? Do we look forward to a future of technological solutions or one of limited access to resources? And how will our dialogues shape our material future? 'Of Earth, For Earth' is a collection of thought pieces from those close to the mining industry and those further from the mining industry; from academics in multiple disciplines, mining and social practitioners, artists and writers. We do not aim to disseminate information, but to listen to different perspectives and experiences, to find a way forwards through dialogue to a sustainable future based on the materials 'Of Earth, For Earth'.

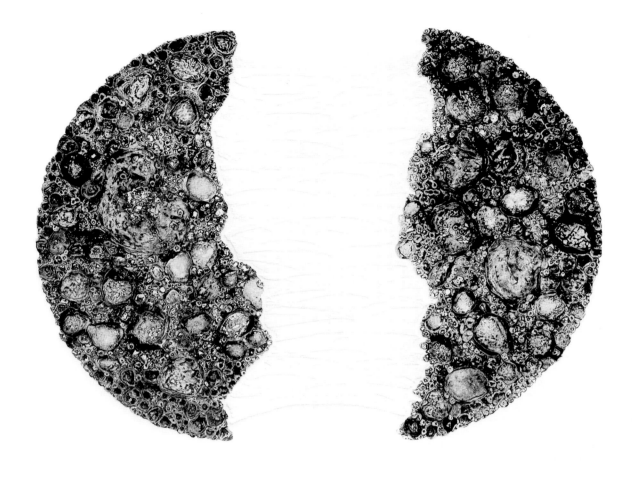

The Art of Material Consumption.

The anthropologist Professor Augustine Fuentes recently stated that 'Human creativity is our ability to move back and forth between the realms of "what is" and " what could be"'. It is the space between the material and the imagination. For artists this space is where we exist. It is where we think, play, learn and ultimately make. The relationship with our chosen materials is a key factor in what we make. The dialogue between the ideas we engage with, the visual stimulus of giving that idea form, and the haptic stimulus of how the material feels and behaves is a balance between uncertainty, meditative action and joy. Within my own practice the nature of materials and the processes I subject them to is key. In order to say something about the world I explore, experiment and then combine those discoveries with the themes I am interested in.

Of late however there has been another conversation between the materials and myself. It is also one that I am not alone in having. Amongst so many of my peers the talk has been about the impact of our choice of materials on the planet. Which has less environmental impact: oil or acrylic paints? What are the replacements for heavy metal pigment colours actually made from? What about the bleaching of paper and canvas fibres or the glues in supposedly sustainable supports? That's just the painters, it's even worse if you're a sculptor dealing with plaster, resins, silicon, fillers and glass fibre.

In this respect the open call from the IMP@CT project seemed timely. My proposal for Of Earth, For Earth was, I thought, a simple one. Create a new work that visually represents my response to humanity's consumption of the Earth's resources. Record every physical material I used in its creation and then cost, research and write something about how each material is sourced and its value. However the studio diary I kept and the subsequent research were a Pandora's box that when opened asked serious questions about what, how and even why I make art.

Are we, despite our emotional investment in the cause of the global climate crisis, as guilty of the disconnect between the means of production and our choice of materials as those who do not realise the true cost of a pork chop or a smart phone? Is it really justifiable to make work with an ecological message if you don't actually ameliorate the footprint of that work? Is an artwork even a valid use of resources at all? A good friend recently stated that 'the decision about which works not to make are often more important than the ones you do'. So does any opinion of mine, any heartfelt reaction to the world I live in really mitigate the material consumption used to speak about it?

The problem feels cyclical. We need to keep making work now that responds to the world as we see it. If public pressure on government and industry from those of us who feel these issues most keenly grows, change will come. This should lead to new environmentally safer materials for artists to appropriate. But the material choices we make in our studios have an impact today, and tomorrow and for decades after. How should we respond in our own practice now? It takes an artist long enough to find their own voice, so changing direction is no simple matter. Can I still have a strong individual voice if many of the materials I have had at my disposal are now morally unjustifiable?

We cannot avoid being consumers of resources. Nor can we refrain from being producers of waste. However, we can seriously manage the manner in which we conduct these material interactions. Not making art is for me not an option. The form it takes will undoubtedly have to change but for me art is not disposable, creativity is a core tenet of what makes us human. Along with everyone else we must scrutinize the environmental impact of our material consumption.

In her opening to this book Dr Kate Moore asked how our dialogues would shape our material future? 'This is not about the dissemination of information. This is a narrative of individuals.' I would suggest that it is so much more than that. For those who move between the realms of 'what is' and 'what could be' it is a collective narrative and creativity our universal voice. For those who speak so much more eloquently through visual language than on any podium this is how we start that dialogue.

To expand upon Ai Weiwei's famous saying perhaps 'Every thing is art' and 'Everything is politics' but Earth is everything.

Opposite: 'Working the Earth' 490mm x 600mm, Collagraph with blind embossing, using hand mined Bideford Black pigment on Somerset Velvet paper, 2020 Dan Pyne.

Overleaf: 'Costing the Earth' 1.2m x 1.54m mixed media on panel, 2020 Dan Pyne.

alternative to expanded plastic packaging. Markets for new packaging materials with mechanical properties very similar to those of some expanded plastic have shown huge growth. Known commercially as Paper-Foam this can be coated in corn protein to protect it from liquids and grease but is biodegradable and can be recycled with ordinary paper.

POLYURETHANE. Use: 1.2kg as foam. Cost £5.59 per 0.75kg. 3.5kg as resin. Cost £14.00 per kg. Average global price: As liquid resins US$3,000-3,500 per ton.

Polyurethane plastics exist in a variety of forms, including flexible foams, rigid foams, chemical-resistant coatings, speciality adhesives and sealants, and elastomers. Polyurethanes are used in the manufacture of high-resilience foam seating, rigid foam insulation panels, micro-cellular foam seals and gaskets, durable elastomeric wheels and tires (such as roller coaster, escalator, shopping cart, elevator, and skateboard wheels), automotive suspension parts, electrical potting compounds, high performance adhesives, surface coatings and sealants, synthetic fibres (e.g, Spandex), carpet underlay, condoms, and hoses.
As solids polyurethanes are inert however the liquid resin blends and isocyanates contain hazardous or regulated components. Isocyanates are known skin and respiratory sensitisers. Additionally, amines, glycols, and phosphate present in spray polyurethane foams present risks. The risks of isocyanates was brought to the world's attention with the Bhopal disaster, which caused the death of nearly 4000 people. The release of methyl isocyanate was the cause of this disaster.
The increasing interest in sustainable "green" products raised interest in cleaner biobased and isocyanate-free polyurethanes. There is potential to derive polyols from vegetable oils such as soybean, cotton seed, neem seed, and castor.

POLYETHYLENE. Use: 4.37m2 as flash spun, high density paper. Cost £1.24 per m2 <600ml as powdered fibre thixotrope. Cost £11.99 per ltr. Average global price: (2019) virgin HDPE granules US$ 600-1260 per ton.

Polyethylene or polythene is the most common plastic. As of 2018, over 100 million tonnes of polyethylene resins were produced annually, accounting for 34% of the total plastics market. The primary use is in packaging, plastic bags, plastic films, geomembranes, containers including bottles, etc. Polyethylene is produced from ethylene, and although ethylene can be produced from renewable resources, it is mainly obtained from petroleum or natural gas.
Both High Density (HDPE) and Low Density Polyethelene (LDPE) can be readily recycled. They can be melted, cooled, and reheated again without significant degradation. The widespread usage of polyethylene poses difficulties for waste management if it is not recycled. Polyethylene, like other synthetic plastics, is not readily biodegradable, and thus accumulates in landfills. However, there are a number of recently discovered species of bacteria and animals that are able to degrade polyethylene.

ACETIC ACID. Use: 500g as distilled vinegar 5% acidity. Cost £0.40 per 568g. Average global price: Bulk acetic acid US$414 per ton.

Acetic acid or ethanoic acid is an organic compound, important as an industrial chemical and reagent. It is used primarily in the production of cellulose acetate for photographic film, polyvinyl acetate for wood glue, and synthetic fibres and fabrics. In households, diluted acetic acid is often used in descaling agents. In the food industry, acetic acid is an acidity regulator and also used as a condiment in the form of vinegar.
Acetic acid is produced both synthetically and by bacterial fermentation. About 75% of acetic acid made for use in the chemical industry is made from methanol (methyl alcohol). The biological route accounts for only about 10% of world production, but it remains important for the production of vinegar because many food purity laws require vinegar used in foods to be of biological origin.
Ascetic acid is key for the production of the lightweight recyclable plastic polyethylene terephthalate (PET) and as the emphasis on recyclable plastics increases this is in turn driving up its importance and production. The global acetic acid market was worth US$ 8.1 billion in 2018.

HYDROGEN PEROXIDE. Use: 300g at 3% solution. Cost £6.99 ltr Average global price: as 50% solution. US$500 per ton.
Hydrogen peroxide is a chemical compound that occurs naturally in biological systems including the human body. Commercially manufactured hydrogen peroxide is used as an oxidiser, bleaching agent, and an antiseptic. Industries include paper manufacture, medicine, water treatment, hydroponic agriculture, aquaculture and as an ingredient in glow sticks. Low concentrations, between 3-6%, are widely available and legal to buy for medical use. However concentrated hydrogen peroxide, is considered extremely hazardous therefore regulated. It has been used as a propellant in rocketry and an explosive.

SODIUM CHLORIDE. Use: 15g as table salt. Cost 35p per 750g Average global price: (2019) rock salt bulk price US $62 per ton and USD $220 for fine open pan salt.

Sodium chloride (NaCl), commonly known as salt, is one of the most abundant minerals on Earth and an essential nutrient for many animals and plants. It is naturally found in seawater and in underground rock formations. It is also one of the largest-volume inorganic raw materials used. Chemical production accounts for over half of all salt use globally. The second-largest consuming segment (15% in 2019) is road salt for deicing.
In manufacturing its uses range from paper, rubber and glass, to chlorine, polyester, household bleach, soaps, detergents and dyes. It is used in metal extraction as an electrolyte to reduce melting point of the minerals It is a valuable food preservative and flavouring agent as well as a mineral in animal diets.
Sodium chloride is also important medically in IV drips to alleviate dehydration, inhalations, saline flushes, eye drops and for cleaning wounds. Salt is available in most countries of the world, and because of its low price, producers are in most cases restricted to regional markets. However, more than 10% of all salt produced and consumed in the world is traded over large distances. The primary exporters are Australia, India, and Mexico and Chile.

GYPSUM. Use: 1.95kg as polymer plasters. Cost £38.22 for 25kg. Average global price: (2019) raw gypsum US $8.00 per ton.

Gypsum is a soft sulphate mineral composed of calcium sulphate dihydrate. It naturally occurs as beds or nodular masses up to a few metres thick and is the product of the evaporation of seawater. Synthetic gypsum (calcium sulphate) may also be derived as a by product of certain industrial processes. The most important is flue gas desulphurisation (FGD), a process that removes sulphur dioxide from the flue gases at coal- fired power stations. Gypsum is used primarily in the manufacture of building products, plasters, plasterboard and cement. High purity natural gypsum is also used to produce specialist plasters, for the ceramics industry and for surgical and dental work. Small quantities of high purity gypsum are also used in confectionary, food, the brewing industry, pharmaceuticals, in sugar beet refining, in agriculture as a soil modifier, as cat litter and as an oil absorbent.
A major advantage of gypsum is that it is infinitely recyclable for re use in plaster board. Unfortunately this gypsum is not currently recycled from demolition sites in the UK.

QUARTZ. Use: 340g as fine and medium ground filler powder. Cost £6.04 per kg. Average global price: as raw silica rock $60-$80 per metric ton.

Quartz is the most abundant and widely distributed mineral found at Earth's surface. It is plentiful in all parts of the world. In its numerous forms it is a key component in glass making, abrasives, foundry sand, refractory bricks, hydraulic fracturing, gemstones, and as a filler in the manufacture of rubber, paint, and putty. It is also used as fine recreational sand in volleyball courts, baseball fields, children's sand pits and manufactured beaches.

Billions of pure industrially grown of quartz crystals are used to make oscillators for watches, clocks, radios, televisions, computers, cell phones, electronic meters, and GPS equipment. A wide variety of uses have also been developed for optical-grade quartz crystals. They are used to make specialized lenses, windows and filters used in lasers, microscopes, telescopes, electronic sensors, and scientific instruments.

Dan Pyne.
Costing The Earth, 2020
Mixed media on panel
or to be more precise...

- - -

TIMBER. Use: 0.0038m3 as planed softwood. Cost £11.93
Average global price: (2019) softwood sawlogs US$54.54 per m3.

About 30% of the global total land area is covered by forests so theoretically substantial amounts of wood can be harvested without depleting or degrading forest resources. The environmental impact of timber use is not straightforward as large amounts of energy is needed to dry and process it. There has been a rapid growth in sustainable commercial, higher density coniferous forestry. This is exemplified by the fact that Asia Oceania, North America and Europe have, combined, extracted over 75,000 Mm3 of logs since 1990 (which accounts for just under three-quarters of global production) and yet have increased combined forest cover by 1Mha/yr. This, is in stark contrast to the management of forest resources in the tropical, developing regions of South America, Africa, and south-east Asia, which account for over 80% of global forest losses through degradation and deforestation

PLYWOOD. Use: 1.48 m2 x 6mm (0.01m3) as panel. Cost £10.75
Average global price: (2020) $5.65 per sheet.

Plywood is manufactured by laminating softwood and hardwood veneers with glues primarily made from urea-formaldehyde or phenol-formaldehyde resins. It is manufactured worldwide with Asia accounting for around 62% of global production (2016). While European and North American wood supplies are normally from Certificated (FSC, PEFC) sustainably managed forestry, most others are not.
Formaldehide is a key petrochemical derivative with wide applications. In addition to its use in the production of resins it is used as a biocide, disinfectant, slow release fertiliser and preservative. Formaldehyde is a suspected carcinogenic, mutagenic and teratogenic compound. It can be released into atmosphere or water resources during production processes and frequently appears as an indoor pollutant because of the way urea-formaldehyde resins break down. Increasing concerns over its use have led to more stringent testing methods and emissions classifications being instituted by the EU, USA and Japan. Industries are being pushed towards replacing urea-formaldehyde resins with low or formaldehyde free adhesives but as they are more expensive to produce it is still prevalent.

IRON. Use: 430g as cast iron powder. Cost £12.36 per kg.
Average global price: Bulk iron ore US$92.2 per ton.

CARBON STEEL. Use: 92g as fixings. Cost £6.85. Average global price: (2019) Steel US$685 per ton. Iron ore US$92.2 per ton

Iron ore is the second largest globally traded commodity, behind crude oil, but converting it to into steel is an enormously energy intensive process.
It takes around 770 kg of coal to produce 1 metric tonne of steel using the basic oxygen (blast) furnaces that currently produce about 74% of the world's steel. A further 25% of steel is produced in Electric Arc Furnaces. In 2018 it is estimated that the global steel industry used about 2 billion tonnes of iron ore, 1 billion tonnes of metallurgical coal and 575 million tonnes of recycled steel to produce about 1.8 billion tonnes of crude steel. 2.1 tonnes of CO_2 is emitted from the raw materials used to produce every tonne of steel.

ZINC CHROMATE. Use: <1g as passivating coatings on steel fixings. Cost <£0.01 Average global price: chromium metal price (2016) $7,150 per ton. Chromite ore price $187 per ton.

Zinc is commonly used alone or in conjunction with other coatings as a corrosion inhibitor for fastenings and other metal components.
Chromium compounds, in either the chromium III (triavelent) or chromium VI (hexavelent) forms, are used for chrome plating, the manufacture of

dyes and pigments, leather and wood preservation, and in smaller amounts for drilling, textiles, ink toner, and toner for copying machines. However, chromium VI is both toxic and carcinogenic, thus highly regulated. Europe restricts hexavalent chromium, but has no restrictions on the less hazardous trivalent chromium which is now widely used as a replacement. Stainless steel and electroplating together comprise 85% of chromium's commercial use.

POLYVINYL ACETATE Use:150g as adhesive & surface primer. Cost £3 per kg. Average global price: Vinyl Acetate monomer (2016) US $820 per ton.

Polyvinyl acetate (PVAc) is a synthetic resin derived from ethylene gas, an key industrial organic chemical. Ethylene production is dependant on crude oil and natural gas. The water emulsions of polyvinyl acetate are commonly used in paints and adhesives, construction, paper and cloth, due to their low cost and resistance against biodegradation. The Global market for polymer emulsions is estimated at 29.8billion (2018). This is rising driven by increased global consumption, demand for low VOC paints, coatings, adhesives and eco products needed to comply with tighter government safety regulations.
The largest producers of vinyl acetate monomers are the US, China, and Western Europe with the Asia Pacific region consuming over 60%.

ACRYLIC CO-POLYMER. Use: 1.35kg as polymer plaster additives, <100g as Co-polymer binder in acrylic paints Cost £7.00 per kg
Average global price: (2019) US$ 1500-5000 per ton.

A polymer emulsion is the suspension of polymers in a liquid. As the liquid evaporates, the suspended polymer solids come closer together until they touch and combine to form larger chains and eventually a film. A paint can be made by pigmenting a polymer emulsion. The type of polymer used determines the type of paint or medium, acrylic polymers for acrylic paints and vinyl polymers for vinyl paints.
There are numerous acrylic polymer formulas depending on what percentage of the emulsion is composed of resin solids. Most paint manufacturers blend the various polymers to obtain the best working characteristics. A blend of more than one polymer is referred to as a co-polymer; virtually all artist acrylic and vinyl paints are co-polymers
In more recent developments acrylic polymers have been added to hard gypsum plasters to form polymer-modified plaster. This is significantly tougher and less porous than normal plaster. So much so that it has been compared to resin. They are non-toxic, non-flammable, with no smell.

ACRYLIC GESSO. Use: 50g as primer Cost £12.95 per kg.
Average global price: unavailable as bulk price

Most commercially available acrylic gessoes contain a mix of acrylic polymers, calcium carbonate, titanium dioxide pigment, 1-5% monopropolinglycol and ammonia or formaldehyde.

Glycol ethers are used as solvents in resins and paints, and can be found in cleaning compounds and cosmetics. Exposure to high levels of glycol ethers can result in narcosis, pulmonary edema, upper respiratory tract irritation and severe liver and kidney damage.

PAPER. Use: 4.98m2 100gsm coated paper. Cost £12.95 per ream
Average global price: Bulk 100gsm coated US$500-900 per ton.

The environmental impact of paper is significant, which has led to changes in industry and behaviour at both business and consumer levels. Global environmental issues such as air and water pollution, climate change, overflowing landfills and deforestation have all lead to increased government regulations. There is now a trend towards sustainability in the pulp and paper industry as it moves to reduce clear cutting, water use, greenhouse gas emissions, fossil fuel consumption and clean up its impacts on local water supplies and air pollution.
The processing of recycled paper is more costly and time consuming than to create virgin paper. Most recycled paper is therefore priced higher than freshly made paper, and this is a large factor for consumers.
Manufacturers have also started to see paper as an environmentally friendly

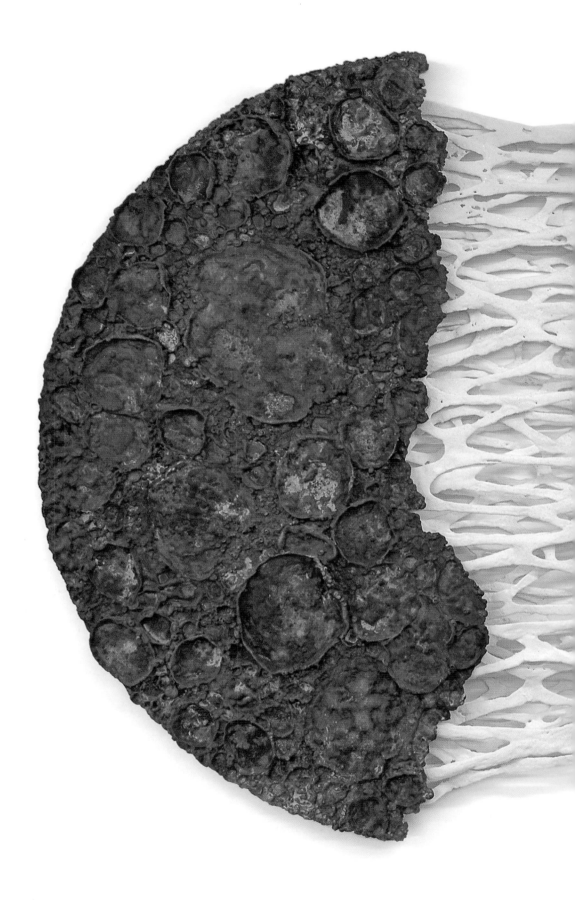

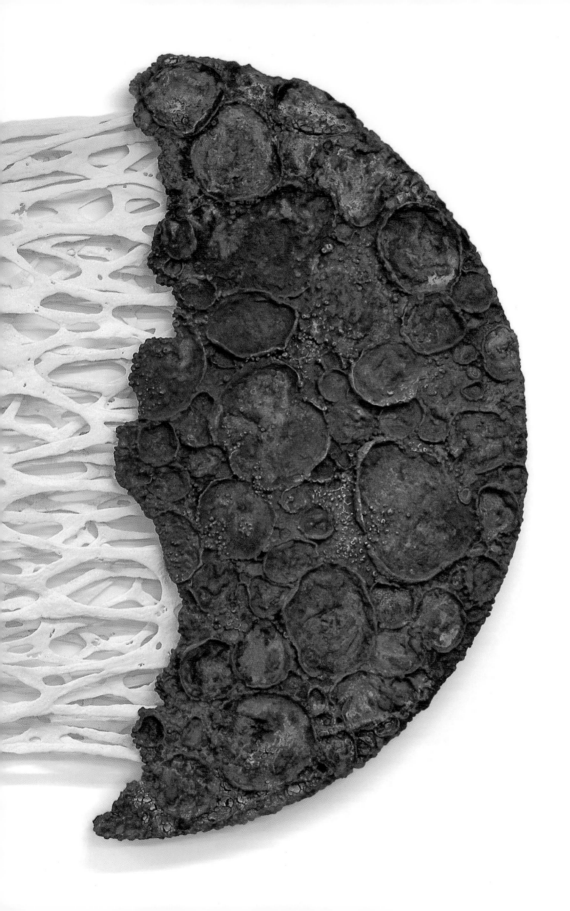

wavelength but also the wavelengths of ultraviolet rays and other colours lower in the visible spectrum. As a result, your eye perceives a far more intense colour. Where a clean, bright conventional colour is able to reflect a maximum of 90% of a colour present in the spectrum, a fluorescent colour can reflect as much as 200% to 300%.

As with all cadmium compounds elements in these pigments are today considered toxic and dry pigments need careful handling.

CERULEAN BLUE. Use: <20ml as water soluble acrylic. Cost £7.30 per 75ml. 400ml as solvent based nitro-acrylic spray. £5.95 per 400ml. Average global price: Cerulean Blue Hue pigment. US$46 per kg. Finest Cerulean Blue pigment US$ 240.00 per kg.

Classicaly the word caerulum was used to describe green blue pigments, particularly mixtures of copper and cobaltous oxides, like azurite and smalt. Early attempts to create sky blue colours were often less than satisfactory due to a limited saturation and the tendency to discolour in reaction with other pigments.

The primary chemical constituent of cerulean pigment is cobalt stannate. The precise hue of the pigment is dependent on a variable silicate component. True cerulean is an expensive pigment so variation often called Cerulean Blue Hue are manufactured. These can either contain cobalt chromite blue-green spinel or be based on the much more recent discovery (1935) phthalocyanine blue mixed with titanium white.

COBALT BLUE HUE. Use: <20ml as water soluble acrylic. Cost £5 per 60ml. 400ml as solvent based nitro-acrylic spray. £5.95 per 400ml. Average global price: Cobalt pigment US$ 219 per kg.

Smalt, a pigment made from cobalt blue glass has been known since the Middle Ages and in impure forms cobalt blues have long been used in Chinese porcelain. Refined Cobalt Blue pigment was only established in the nineteenth century. It is made by sintering cobalt(II) oxide with alumina at 1200°C.

Cobalt ore is difficult and dangerous to extract from the ground. It is primarily extracted as a by-product of nickel and copper ores and mined in large quantities in the Democratic Republic of the Congo, China, Russia, Canada, and Australia. The dust, which can contain toxic elements like cobalt, arsenic, sulphur and uranium is released during the mining process is extremely hazardous to mine workers and contaminates water sources. The cobalt is also used to produce lithium-ion batteries used to power electric cars, laptops and smart phones so of extremely high value.

As with Cerulean Blue approximated alternatives or Cobalt "Hues," are plentiful. They generally contain a mix of Ultramarine or Pthalo Blue with a white.

ULTRAMARINE. Use: <20ml as water soluble acrylic. Cost £5 per 60ml. 800ml as solvent based nitro-acrylic spray. £5.95 per 400ml. Average global price: synthetic ultramarine US$ 4-10 per kg. Finest ground lapis lazuli ultramarine US$30,643 per kg

Traditional Ultramarine is one of the most complex mineral pigments. It is a mineralized limestone containing a blue cubic mineral called lazurite, the major component of the semi precious stone lapis lazuli.

Due to the rarity of lapis lazuli and the complicated process needed to extract it ultramarine was traditionally the finest and most expensive blue available to artists. During the renaissance It was reserved for the robes of the Virgin Mary. It remained an extremely expensive pigment until a synthetic ultramarine was invented in 1826.

The synthetic version, often called French Ultramarine, is chemically identical to the original pigment. However synthetic ultramarine is a more vivid blue since the particles in synthetic ultramarine are smaller and more uniform and therefore diffuse light more evenly.

TITANIUM DIOXIDE. Use: <25g as water soluble acrylic paint. Cost £6.00 per 205g. 50g as polyurethane colourant. Cost £6.00 per50g. Average global price: (2018) as standard rutile TiO2 pigment US$2,600-3,000 per ton.

Titanium dioxide is a modern synthetic white pigment with high opacity

and permanence, low toxicity, and reasonable cost. It has therefore eclipsed most other traditional white pigments. Titanium dioxide is produced from either ilmenite, rutile or titanium slag.

Titanium dioxide is the most important material used by the paint industry for whiteness and opacity. It has the highest refractive index of any material known to man, greater even than diamond. Besides paint it is used in coatings, adhesives, paper, plastics and rubber, printing inks, coated fabrics and textiles, as well as ceramics, floor coverings, roofing materials, cosmetics, toothpaste, soap, water treatment agents, pharmaceuticals, and food colorants. The paint industry consumed nearly 3.5 million tonnes of TiO2 in 2019, accounting for 55% of total world consumption. The global Titanium Dioxide (TiO2) market was US$ 16.2 billion in 2018.

IRON OXIDE. Use: 55g as yellow ochre pigment. Cost £0.00 collected by the artist. Average global price: (2019) price finished iron oxide (Fe2O3) pigment US$1270 per ton.

Iron oxides are naturally occurring compounds composed of iron and oxygen. There are sixteen known iron oxides and oxyhydroxides and all are widespread in nature. They are commercially used as iron ores, pigments, catalysts, and in thermite. Oxides are also used in a variety of filters, inductors and transformers in electronic home appliances and industrial equipment, as well as in flexible magnets, generators, loudspeakers and electric car motors.

Iron oxides are inexpensive and durable pigments in paints, coatings and coloured concretes. They may be naturally occurring or synthetic and form the majority of the "earth colours" in the yellow/orange/red/brown/black range. Ochres, siennas and some umbers are invariably derived from limonite while magnetite provides a black iron oxide pigment.

MANGANESE DIOXIDE. Use: 60g as Raw Umber pigment. Cost £0.00 collected by the artist. Average global price: (2019) as the commercial ground pigment Umber £4.80 per 100g as Pyrolusite ore averaged US$4.50 per ton.

Manganese is the twelfth most abundant element and the fifth most abundant metal, comprising about 0.1% of the Earth's crust but does not occur naturally in its native state as a base metal. It is found in over 100 minerals, the most common of which is Pyrolusite, containing 60-63% manganese. More than 80% of the high-grade ore is mined in South Africa but China, Australia, Brazil, Gabon, India, and Ukraine are major producers. Large manganese occurrences exist on deep ocean floors.

Manganese dioxide is used in the production of steel and metal alloys, glass, batteries, electronics, fertilizers, animal feed, water treatment, and as a pigment for paints, bricks, glass, textiles, and tiles. In artists' paints it forms the basis of the Umber family of brown pigments.

BIDEFORD BLACK. Use 80g as raw pigment. Cost £0.00 collected by the artist. Average global price: Not commercially available.

Bideford black is a soft, carbon based mineral containing silica and alumina, with the black colouration created by the carbon. It occurs as a seam of harder anthracite coal and organic rich shales. The deposits were formed 350 million years ago during the Carboniferous period. Bideford black is believed unique in that it contains a dominance of the lignin of tree ferns rather than the mix of stem, spores, bark and leaf matter normally associated with coal deposits. It was mined locally for fuel for steam engines and lime-burning and commercially as a pigment for paints, dyes, rust proofing for iron clad ships, tank camouflage and as a mascara ingredient. Commercial mining of the Bideford Black pigment ended in 1969.

- - -

Approximate cost of materials: £229.61
Price of artwork: £2,700
Total cost to the Earth of extracting raw materials?

COLLOIDAL SILICA. Use: 25g as 40 - 80 μm glass microspheres. Cost £15.99 per 120g. Average global price: as umed silicone dioxide US$500-$900 per ton.

Silica is one of the most complex and abundant families of materials, existing as a compound of several minerals and as synthetic product.
It most commonly found in nature as quartz, in the shells of various marine organisms, and as the major constituent of some sands. Synthetic examples include fused quartz, fumed silica, silica gel, and aerogels. Silicon dioxide is mostly obtained by mining, (including sand mining) and by the purification of quartz.
About 95% of the commercial use of silicon dioxide occurs in the construction industry in the production of concrete. Silica is a common additive in food production, as a flow or anti-caking agent in powdered foods. It is used to adsorb water in hygroscopic applications like silica-gel. Colloidal silica is used as a wine, beer, and juice fining and as a fine abrasive in toothpaste, and household cleaning products. In a fibrous form it is useful as a high-temperature thermal protection fabric.
Silica microspheres are increasingly used as lightweight fillers in resin, paint, soap and makeup as they improve flow and oil absorption. They are also an inorganic support used in the extraction of DNA and RNA. Ingested silica is essentially nontoxic but there are major health risks associated with, airborne, crystalline silica particles, such as lung cancer, silicosis, chronic obstructive pulmonary disease, and kidney disease.

KAOLINITE. Use: 80g as powdered china clay. Average global price: (2017) US$150 per ton.

Kaolinite is a soft, earthy, usually white, mineral (dioctahedral phyllosilicate clay), produced by the chemical weathering of aluminium silicate minerals like feldspar. It is mined as kaolin, locally in Cornwall as well as Malaysia, Pakistan, Vietnam, Uzbekistan, Brazil, Bulgaria, Bangladesh, France, Iran, Germany, India, Australia, South Korea, China, Czech Republic, Spain, South Africa, and United States. The largest single producer of raw Kaolin is Uzbekistan. The global kaolin market was valued at US$4.36 billion in 2019.

Approximately 40% of global kaolin use is as a filler and coating in paper production. 30% is used in the ceramic industry for the manufacture of whiteware and porcelain. It is used in paints, inks and industrial coatings, as a strengthening agent in rubber and plastics and in the manufacture of glass fibre. It can be found in electrical cabling as insulation and in modified forms as a concrete additive. Medically kaolin is known for its capabilities to induce and accelerate blood clotting. Kaolin also has uses in agriculture as a carrier for fertilisers and pesticides and as a seed coating.

CALCIUM CARBONATE. Use: 520g as fine marble dust. Cost £3.42 per kg. Average global price: unrefined ground calcium carbonate (depending on grade) US$50-300 per ton.

Calcium carbonate is a common compound found in rocks particularly as the pure calcium carbonate minerals calcite, aragonite, and vaterite. It is also the predominant compound in the industrially important rocks limestone, chalk, marble and travertine.
It has major industrial applications including: adhesives and sealants, animal feeds, concrete, plasters, asphalt, fertilisers and agricultural liming, food supplements, pharmaceuticals, glass, ceramics, household cleaning products, paints thickeners, surface-coatings, paper, plastics and composites, rubber and elastomers.
The vast majority of calcium carbonate used in industry is extracted by mining or quarrying. Purified forms of calcium carbonate such as for food or pharmaceutical use are further refined from quarried sources (usually marble). The global calcium carbonate market was valued at USD 22.95 billion in 2018.

MICA. Use: 30g as fine flakes. Cost £15 for 500g.
Average global price: fine muscovite flake US$200-$500 per ton.

Mica is a general name for a group of silicate minerals having thin sheet-like or plate-like structure including muscovite, biotite, lepidolite, phlogopite, zinnwaldite and clintonite. It is widely mined as both sheets and flakes.

A synthetic mica is also increasingly being manufactured. Natural mica accounts for 90 per cent of the total mica market.
While scrap and flake mica is produced all over the world, the most valuable high-quality sheet mica comes from China, India, Brazil and Russia. 2010 figures show it sold for as much as US$2,000 per kilogram.
Mica is valued for its strength as a thermal and electrical insulator. It is used in electrical components: capacitors, transformers, semi-conductors, circuit boards, electrical cables, insulation and lithium ion batteries. Sheet mica is found in aerospace components, missile and radar systems, laser devices and medical electronics. Ground mica is used in construction and plastics industries as both fillers, lightweight extenders and decorative finishes and by the oil industry as an additive in drilling muds.
It is used in paints as a pigment extender, a colour brightener and for pearlescent and metallic finishes. The reflective properties of ground mica also make it a valued ingredient in cosmetic products. Mica is added to eye shadow, lipsticks, blushes, body glitter, nail polish, shampoo, toothpaste and other products to create pearlescent effects
The electronics industry was the largest purchaser of mica in 2015 (26 per cent), followed by paints and coatings (24 per cent), construction (20 per cent) and cosmetics (18 per cent).
There are serious global concerns regarding mica mining. India is the largest producer of fine sheet mica but experts estimate that approximately 70% of the mica production there is the result of illegal mining. Dutch campaign group Centre for Research on Multinational Corporations estimates up to 20,000 children are involved in mica mining in India, some as young as 6. Child labour is also employed in mica collection in Madagascar and is suspected in China, Brazil, Pakistan and Sudan.

PIGMENTS

LEMON YELLOW. Use: <20ml as as water soluble acrylic. Cost £6.30 per 150ml. Average global price: True Barium Chromate Lemon Yellow US$72 per kg.

CADMIUM YELLOW. Use: <10ml as water soluble acrylic. Cost £9.80 per 75ml. Average global price: Cadmium Deep Yellow raw pigment US$165 per kg.

Cadmium pigments are a class of pigments that have cadmium sulfide as the key chemical component. If selenium is added in place of some sulphur the yellow shade can be deepened through orange to red. Much of the cadmium produced worldwide has been used in the production of rechargeable nickel-cadmium batteries, but of the remaining consumption roughly half goes into to production of cadmium pigments
The pigment known commonly as "Lemon Yellow" can consist of barium chromate, strontium chromate, or a mixture of lead chromate with lead sulphate. The preparation was described by French chemist Nicolas Louis Vauquelin in 1809, but some years seem to have passed before it was produced commercially. During the nineteenth century, the name 'Lemon Yellow' was primarily used for barium chromate, strontium chromate or a mixture of the two. It is a stable yellow with low opacity.
As with so many 19th century pigments the toxicity of the chromate and cadmium sulfide compounds used in these original yellows has forced manufacturers to find alternatives. In 2015 the European Chemicals agency granted a reprieve for the specific use of cadmium in artists colours for historic applications, however the chromates are being replaced. Apart from a few low volume hand crafted paint manufacturers most commercial Lemon Yellows now derive from a family of organic compounds called azo dyes or from nickel titanate derived from by heating titanium oxides, nickel, and antimony together. Arylide Yellow has become the standard bright Lemon Yellow. While Nickel Titanate Yellow is paler creamier and closer to the original Barium Chromate yellow.

FLUORESCENT RED. <40ml as water soluble acrylic. Cost £3.50 per 60ml. Average global price: Daylight fluorescent pigment US$159 per kg.

Fluorescent colours contain inorganic fluorescent pigments such as zinc sulphide and zinc cadmium sulphide that use a larger amount of both the visible spectrum and the lower wavelengths compared to conventional colours. They not only absorb and convert light energy of the dominant

Should the cost
of avoiding disasters
be embedded in
every cell phone,
laptop, pot, pen,
lamp?

Carlos Petter.
Mining Disasters As Death Foretold?

In his brilliant book "Chronicle of a death foretold", written in 1981, the Colombian author Gabriel García Marquez, tells the story of a murder, Santiago Nasar's murder. Paradoxically, it puts us in a situation of latent suffering, because in the first chapter he tells us that the main character dies. We readers are held in anticipation to the very last page, hoping that some magical or divine intervention occurs to change Santiago's destiny. Against all our hopes, the main character ends up really dying!

Mining disasters are also, in their own way, a chronicle of a death foretold. We - technicians, engineers, managers, stockholders, stakeholders - sincerely do not want them to happen, but they do happen and, worse, they continue to happen. Unhappily, cost management is one of the main actors (Armstrong et al., 2019). As examples, those dam failures in Canada (2014) and Brazil (2015 and 2019), were colossal human and environmental tragedies! Human and environmental tragedies had already happened in the last century and this merely accelerated in the 21st century (https://worldminetailingsfailures.org/). They have happened throughout the history of mining, past and present, and there is no guarantee that they will not happen in the future, more likely the contrary.

Should we stop mining activity? My answer: this would be very difficult! Only with quantum leaps of knowledge in the field of energy would that be possible. If we could generate clean, cheap and abundant energy, the closed cycle of urban mining via recycling will be within our reach. Then, we can think about breaking the spiral that has led us to mine more in the past 30 years than in all of humanity's past history. In the meantime, let's get back to reality...

The time has come for us to reflect on those who work within mining, and on how we teach and communicate. With ourselves, our children, our parents, neighbours, students, all those who are and will be the consumers, entrepreneurs and technicians of today and tomorrow. We need to re-think the valuation process. Should the cost of avoiding disasters be embedded in every cell phone, laptop, pot, pen, lamp? This is one of the challenges to which I am devoting much of my energy today, knowing that, sometime in the near or not so near future I will not be here anymore. For this reason, I unconditionally support projects like IMP@CT for a sustainable future based on the materials 'Of Earth, For Earth'.

Alan Smith & Louise K Wilson.
72 Hours Chthonic.

"Neither the sea nor the forest so lends itself to the substantiation of the supernatural as does the mine" Agricola, De Re Metallica.

The curiosity to inhabit the 'underland' was the starting point for this durational arts project. On the 20th April 2017 Alan Smith, John Bowers, Louise K Wilson and Peter Mathews met and prepared to enter Ballrooom Flats, Smallcleugh mine in Nenthead, Cumbria. Journeying in was exhausting, relentless and initially dispiriting, with movement slowed from having to carry water, food, clothing, bedding and technologies essential for their 72 hour duration. After the 6 hour descent, they arrived depleted, fell into a deep sleep and awoke to total darkness. Chthonic had begun.

Since moving to Allenheads in 1994, Smith has explored the post-mining landscape of the North Pennines. Travelling alone into the mines, he would select a place and sit in complete darkness and listen and 'sense' the surroundings. Later he wanted to 'share' the mine with others, to gauge reactions and responses to the extraordinary environment. His first collective piece was Parameter (2009) in which a group of participants spent 24 hours in the mine in silence. Chthonic was subsequently conceived from a desire to see what a longer duration and a more open process might yield: "We will take as our primary set of tools our senses, our knowledge, imaginations and time to ruminate and on return to terra firma we will assay our treasured thoughts and alchemically strive to disclose another way of making" (Alan Smith 2016). John Bowers is a sound artist improvising with electronic, digital, acoustic and electro-mechanical devices and Louise K Wilson has an on-going interest in the acoustics of ruins and of memory in relation to sound. Peter Matthews was interested in investigating possibilities for generating electricity from the limited water flow in the mine. He sought to fabricate a hydroelectric turbine sufficient for the production of works.
The trope of extraction (Smith's term) was devised and offered as a conceptual term to help shape creative and practical thinking, just as the miners did, we would bring materials out into the light for processing and distribution, we would extract, process and refine our experiences.

For Chthonic there was no project plan other than to dwell underground. But we were mindful of an unstated understanding to look after each other and attempt to do some sort of unspecified work. We soon noticed how perceptions were quickly affected by living in the dark environment: our dreams became vivid, and auditory and visual hallucinations were felt.

Darkness and silence quickly gets 'filled in'. Auditory and visual illusions are common for those who travel through such reduced sensory spaces, and the four quickly became accustomed to the everydayness of hallucination. They came to talk increasingly of 'the fifth person'. Most compelling certainly was the imagined vocals heard in particular parts of the mine. Standing not far in from the 'adit' entrance for example, 'voices' akin to the babble of young children could be heard emanating from the interior. Once suggested by Smith it was impossible not to hear the 'children'. The real sounds of snoring and breathing filled the Ballroom with rhythms that Smith described as comforting and reassuring.
Exiting the mine was a far swifter journey and the desire to hasten the last few metres, first smelling the fresh air nearing the adit, then lying on the ground, the sublime feeling of daylight on the skin, was exquisite.

Chthonic can be articulated as a starting point for an unfolding empirical aesthetic program and could finally, be summed up as a developing and extended infrastructure, as a resource to help think and make, to draw on both separately and collaboratively.

"It might seem a curious thing to consider but when looking at the goal of mining, it is the end of process that is sought after and of value. What stays with me from Chthonic as the most valuable commodity is the shared experience of spending those 72 hours underground, sleeping and waking in absolute darkness, listening to my colleagues breathing and

Image from '72 Hours Chthonic', 2017 Photo: Nat Wilkins.

wondering why it is I can make myself believe in the impossibility of seeing the corners of the Ballroom when there is absolutely no light. It's those experiences, memories and my contemplation of them that continues to feed my practice and is perhaps the artwork, rather than the product". (Smith)

The name 'Ballroom' comes from an event that apparently took place on September the 2nd 1901, when twenty-eight Masonic members travelled into the mine for a banquet and dance (though sadly no documentary evidence is readily available to substantiate this).

Dylan McFarlane.
The Portrayal Of Mining In Society.

Society and mining have depended on one another for thousands of years, since the earliest extraction of clay or stone for tools and weapons as early as 4,000 B.C. Today, our entire built civilization all starts with a hole in the ground, from the clay bricks forming the walls around us, to the copper wires and other rare metals which run through and power our electronic lives. And although society is increasingly dependent on mining, the portrayal of mining is that it is dirty, exploitative, destructive and dangerous. The popular mining headlines of 2019-2020 have fixated on dire stories such as the death of children in Congolese cobalt mines, dam collapses destroying rivers, or the "barbarism" of artisanal and small-scale mining.

Some of that narrative is true and justified. Mining, by definition, exploits the world beneath our feet. In its immediate footprint, mining destroys the living biotic world above ground, breaking into the soil and rock below. Mining is a dangerous, complex endeavour, and humans are prone to accidents, causing pollution. In recent years, as the scale of industrial mining has grown exponentially, the risks have increased, and some notable disasters have caused incredible harm and damage to people and planet.

But modern mining has also evolved with societal demands, responding to criticism from social activists, protestors, civil society, and local communities. Today, most industrial mining operations have higher standards for safety and environment than many farming operations, and deliver higher economic benefits, albeit being "single-use" opportunities. Modern mines must meet stringent criteria from government regulators, but the most significant pressure is from the investment community. To attract the tens or hundreds of millions of dollars needed to build and operate a mine, companies must attract the attention of financiers, who now rank mining projects strongly on "ESG" credentials, which stands for Environmental, Social, and Governance. If a company cannot gain the social acceptance of a community, investors will simply withdraw their money.

Yet the media love a sensational story. And often, the most sensational story is that of the artisanal and small-scale mining sector, "ASM". ASM provides approximately 20% of the world's minerals, characterized by labour-intensive methods, basic equipment, and hazardous working conditions. These hidden producers of the world, often informal and unlicensed, attract the media's attention because the images are striking – a poor African woman with a child tied to her back descending underground in order to scramble a few pence worth of some low value mica flakes. The issue of children working in mines has been exploited by tabloids and even respectable periodicals, but is often orchestrated, set-up, and choreographed in order to tell a narrative that will sell papers, or "clickbait".

The true story of mining is hard to discern, and an accurate evaluation of its impacts on society hidden behind the headlines or public statements. Media conglomerates have political agendas, and as the mining industry continues to be poor at communicating honestly and openly, the vacuum is filled by sensationalism. It shouldn't be this way, nor does it have to be. There are many positive, interesting, empathetic and nuanced portrayals that could be given. Society and the mining sector could be more honest, such as about our dependence on minerals, and about the skeletons in the industry's closet. Mining must strive to improve performance, and it must open more honest, sincere and transparent communication channels. Programmes such as the Initiative for Responsible Mining Assurance (IRMA) act as a "social audit", so to speak, of communities assessing mining operations. Reducing the misinformation spread on social media could perhaps also help. Without these efforts, continued divergence and growing social conflict ensues. It's a paradox that at a time of modern civilization's greatest dependence on mining (a wider range of materials than ever), societal acceptance and the portrayal of mining is at its lowest, most critical juncture. Let's hope it changes.

Adele Rouleau.
Mining: An Industry in Search of a Conscience.

Mining is one of the oldest industries in the world, it has been with humanity from incredibly early, since we have needed the resources of the Earth. However, mineral deposits exist due to the Earth's geological activity over millennia, not because of human ingenuity or creation. Yet mining exists, and we have created systems to extract, process mineral deposits and give them an intrinsic value through our economic system and technological developments. When economic deposits are discovered, the mining industry moves in with all its behemoth of complexity and it tends to elicit an emotional response.

Whoever you ask, whatever their walk in life, will have an opinion on mining. A local community who has been forced to leave their homes and livelihoods due to the construction of a mine will not share the sense of pride in its construction of the mining engineer who has pored over its methodical design. An investor taking a financial gamble on the development of a gold mine will not share a social justice warrior's perspective of mines being bastions of greed and oppression. Mining is an industry which exists at the junction of conflicting interests that, depending on your value system, will lead to a hierarchy of which interest trumps the rest.

Mining is a profitable business and the smell of money grabs the attention of competing parties; mining corporations, governments, local populations. Who will be entitled to the lion's share? The investor gambling on the rise of metal prices, the buccaneering geologist with a love of maps and adventure who caught the first glimpse of a lucrative reserve, the local politician who controls the land and who would like to enjoy the bounty offered by a future multimillion dollar mine? The list of potential beneficiaries could go on, along with the legal fees for the lawyers negotiating the conflicts of rights and entitlements. It is never straightforward, and this is perhaps one of the great beauties of mining law: we will never have a right answer. The fights over land, wealth and taxes are as old as human civilization, it is ultimately a tension between power, natural justice and perspectives; a metaphorical clash of titans.

Mining is a lot more than just technical problems, another rock waiting to be turned into metal. It is a battle for survival: corporate, political, environmental, communal and human. I often get asked if the mining industry needs more regulation and my answer is no, it needs better people. We need engineers who understand the impacts of their actions on local populations, we need politicians who understand what it means to be a public servant, we need charities who understand that mining will continue to exist and does benefit our lives to an immeasurable degree, and we need executives who have a greater vision of a company's role in how it serves wider society and not just an elite few. Laws have an important role to play, however, we need accountability mechanisms, as well as powerhouses that want to be held accountable. This is at its core a call to integrity. There are numerous international mining laws enshrining exemplary clauses on environmental responsibility, benefit sharing and local community empowerment. They are also poorly backed by weak or corrupt government institutions or ultimately undermined by lawyers hired to favour whoever is lining their pockets. As I have heard countless times, "it's just business", "it's just a job", "it's all about what you can get away with", "it is what it is", "it's all about money". I get it, however the mining industry is battling a negative perspective for a reason. Admittedly the situation has changed, many laws we have today were passed in reaction to the negative impacts left by mining, a cry out of "No more", yet we need more than laws to make concrete changes. Mining per se is an objective act which has been a fundamental part of human civilization since the beginning. It is not mining that commits human rights abuses, leaves environmental degradation, prevents wealth from being distributed fairly, it is people.

To end on a simple yet effective quote attributed to the Greek philosopher Plato, "Good people do not need laws to tell them to act responsibly, while bad people will find a way around the laws". The law has a role to play, however it is also a tool used by people to serve an aim, and one always hopes and fights that it will be for a higher purpose.

Josie Purcell.
Elemental Eidos.

Throughout my photographic practice, I aim to find ways to lessen my photographic footprint. This is determined through the use of minimal/non-chemical printing processes or via work that challenges me, and the viewer, to consider alternatives.

Elemental Eidos grew from my series Harena Now which uses an abstract aesthetic to promote conversation about the environmental and humanitarian issues global sand mining is creating. It works on the same principle of using something pleasing to attract attention, and once the engagement is made, sharing the reasons as to why the art has been created.

Drawing on my Harena Now research into Cathodoluminescence, a geological sampling technique, Elemental Eidos has made use of images from this process kindly provided via the IMP@CT team. In addition, some images have been created directly from sand.

The work is mixed up with memories of childhood kaleidoscopes - they provide a touchstone for the elements within, the need to provide and the need to sustain. To create beauty, form and function we need resources.

The question is, how long can we be 'Of Earth, For Earth'? Will we be as multi-faceted as a kaleidoscope, able to change to create a new picture, in our endeavours to secure balance between our wants and modern lifestyles and the availability of natural resources?

'Elemental Eidos, (cathodoluminescence) #2', 20 x 20 inches Lumen/Digital as C-Type Metallic Archive print, 2020 Josie Purcell.

Jack Hirons.
Utopia, It's Down Here Somewhere.

Whilst an interest in the provenance of our food is becoming more popular, there is still a disconnection with other parts of daily life. How did the petrol get to the pump? What do I need to make a phone? How does the internet operate? Or what makes the electricity in our homes? Our daily exchange with materials is very transactional and understandably, it would be a full time job acquiring the information we'd need to make the 'correct' choices.

This leaves us with either complicated decisions to make or often the decision made for us, ultimately making us feel out of control or blindly contradicting ourselves. Even seemingly trivial things for example the line many corporate emails sign off with, 'please consider the environment before printing this email'. Maybe it's better for the environment to print that email and file it, rather than leaving it on a cloud server for the next 20+ years. Technology can improve how we use resources but often it simply changes what we use.

No action has no impact, after all by just existing you are using and changing the environment. Whilst reading this your body is producing carbon dioxide as well as using the energy from what you last ate. What was that? An avocado from Mexico? Or an apple from your garden?

For the work Utopia, it's down here somewhere I've borrowed the existing culture of mining stickers to create a new collection of stickers that combine slogans and pictorial elements to communicate these conflicts we have with the resources beneath the surface of the earth's crust. Some of the stickers refer to specific events in history whilst others purely echo the conflicting narratives in raw material consumption. Collectively the stickers take no particular stance, they remain as confused and conflicted as I am about these issues, which is probably about as confused as humanity is.

Top left: Mining stickers. Above: installation view 'Utopia, It's Down Here Somewhere', 2020 Jack Hirons.

Abandoned mining structure, Olovo, Bosnia, 2017. Photo: Lars Barnewold.

Dominic Roberts.
A Balkan Perspective.

I first went to the Balkans towards the end of the Bosnian civil war as a young second lieutenant in the British Army. For me and many of my Welsh soldiers, some barely eighteen years old, it was our first experience of the devastation of war. We had grown used, inured even, to the horrific images on the television and in the newspapers, but none of the pre-tour training prepared us for the human tragedy that we were to witness. Twenty-five years on I vividly remember the shock of realising that a row of bullet holes along the wall of an abandoned house were not simply collateral damage, but the scene of an execution of an entire family. Whether

the casualties were Bosnian, Croatian or Serb, whether the aggressors were Muslim, Christians or Catholic, it did not matter to us. They were people on the same European continent as we, who lived and for many, died, in fear.

As the summer warmed, we found ourselves exhuming mass graves, often those dug up by wild boar. And then, later that year, we marked out the 'Inter-Entity Boundary Line' agreed by the Dayton Accords, not fully appreciating our role in delineating a new country. It was during this task that a friend of mine and his vehicle's crew were killed by an anti-tank mine in the woods above Glamoč, a site I took his father to years later and one that I return to when I can. A lonely, remote, memorial to three young men killed far from home, doing what I spent most of my military career trying to do, defending the weak and the innocent.

Deploying to Kosovo in 1999, again we saw the human horror of internecine conflict in our supposedly civilised Europe. Again, the bullet hole evidence of a massacre committed. Again, the churches torched and farms abandoned. The burnt-out vehicles and mass graves by the roadside. The old Serbian lady who refused to leave her home of eighty years (where else would she go?), who was murdered the very night that we were ordered to end her twenty-four-hour guard.

Iraq followed, same horrors and many worse. The personal tragedy was still as evident and as upsetting, but without the European context the impact was somehow deadened.

Leaving the Army, I found myself back in the Balkans, but instead of camouflage my uniform was cut of high visibility cloth. Whilst deployed in Bosnia and Kosovo as young soldiers we had not paid much attention to the abandoned industry that surrounded us. But as a fledgling miner, learning my trade in the mineral-rich mountains, I quickly appreciated how deeply the extractive industry influenced the region's prosperity. The locals still reminisce about the glory days of Yugoslavia, particularly of the seventies, when life was good. Everyone was employed. Health, social care and an excellent education were available to all and, bar some oddities of Tito's communist regime, people were free to travel, prosper and enjoy their lives in and amongst the natural beauty that characterises the region.

One still hears the various sides of the stories about the wars, the biased interpretations of history or events. Having read extensively about the region, including the possibly uniquely unbiased works by Noel Malcolm on both Bosnia and Kosovo, I recognise the propaganda but appreciate its genesis. But the simple fact is that the former Yugoslavian states need robust economies. Yes, accession to the European Union is important, as is an end to the ethnic stalemate in Kosovo, but access to employment, tax revenues and the associated state services are what is most valuable to the people of the region. The rural-urban brain drain, the emigration of the brightest youngsters to realise their potential elsewhere, is damaging to the long-term future of these young democracies. Yugoslavia did heavy industry and did it well. Its natural resources can be the very catalyst of this economic revival and are becoming ever more important, as Europe increasingly realises the need to leverage its own mineral resources and not rely on imports. The relationship of mineral resources to conflict is well documented. Whilst they did not particularly influence the breakup of Yugoslavia, they are a dominating factor in its future.

The people of the region are more accepting of mining than any other area I have visited or studied. They are not immune to the environmental and societal impacts associated with the extractive industry, but they very much expect, indeed require, us to do it properly. The much-discussed issue of a social license to operate is not one that troubles our operations.

When I first visited the region twenty-five years ago it was to defend the weak and the innocent. Now I work with them to access the opportunities that are under their very feet.

Olovo, Bosnia Photo: Richard Roethe.

Olga Sidorenko.
Where Communities Welcome Mining.

We are accustomed to images of the mining industry that convey notoriety. Stories about mining in the mass media focus mainly on environmental damage, community and legal disputes, and corruption. We quite often hear about protests against mining in different countries, and calls to implement bans against mining operations.

Considering this, it is hard to imagine that there are many places on Earth where mining is truly welcome. Especially, it is inconceivable that places like this may exist in Europe, where mining has been predominately out of sight for over 40 years and public opposition to it has increased dramatically.

Over the last three years, in my research I have been focusing on some small-scale mining projects in the Balkans. I am not a geologist, but a social science researcher. To me, mining has never been about technologies, new facilities, and methods of extraction but about local people who live near mining sites, their concerns, fears, and expectations relating to mining operations.

While working in the Balkans, reading about the complicated and contested history of the place, listening to people, I was trying to understand why people told me they were eager to see the mines open again. It seemed that they did not have same level of public opposition to mining as in some other European countries. I visited several settlements in Serbia and Bosnia that had emerged largely from a socialist system that was characterized by extensive industrial development. The age of industrial development was also the time of social growth. Citizens of one small village, which now has only two small grocery stores and a tiny two-room medical care facility, told me: "we used to have a swimming pool, a cinema, street lighting, an outpatient clinic with a dispensary, a football club, and a cultural club". By the 1990s, the mining industry had almost shut down, and today these villages are typical rural settlements with a small population, a weak urban system, and lack of opportunities for employment. The population of such villages is elderly, as young people have moved to Belgrade or Sarajevo, or migrated abroad.

The people I spoke to transferred their impressions of the wealth of a mining community that they (or their parents) experienced in their past onto current mining activities. They believe that mining may bring a new wave of development to their regions. It is hard to deny that the return of mining as an economic activity is welcomed there. It is desirable to the local officials, who expect foreign investments in the sector, municipalities who can obtain mining revenues for the depressed regions, and it is also appreciated by the local people. Though they certainly worry about environmental risks associated with mining, overall they believe that mining development will create positive outcomes for society. I was surprised to hear the words of one old man whose house was located just 800 meters from the nearest mine pits. "We all feel that it is our chance for development and better life".

Interviewing people, I noticed that aspects of history that ended in the late 1980s still persist in everyday discussions. And mining sites somehow activate the local memories of the past. I came to Bosnia and Serbia to speak about the expectations of future mining, but instead we spoke a lot about historical perceptions. About the times, more than 40 years ago, when the economic crisis started in Yugoslavia followed by the collapse of the country and continued economic decline. Expectations for the future arising from discussions of current mining operations somehow rely on perceptions of the events that occurred a long time ago.

But to what extent will mining companies and policy-makers be eager to consider these stories and learn from both the negative and positive historical lessons? It is evident that the narratives held by multiple generations can be a source of knowledge for today's decision-making and more sustained and meaningful communication with local citizens. And hopefully, social acceptance of mining granted by the Balkans communities will be spent responsibly by mining companies, ensuring fair distribution of mining benefits and solid environmental protection; while the positive, rational attitude of local citizens to mining as an important industry will remain.

'Orange trees', mixed media on paper, 2020 Dana Finch.

Penda Diallo.
Continuity & Change.

A Conversation With My Grandmother.

It was a calm day. The air under the baobab tree was fresh and clean. The leaves were green and shiny, unlike they have been for the past several years, covered in red dust. I could hear the birds singing, unlike for the past years, when their songs were drowned by the noise of the machines from the mines. When I sat down under the baobab tree, all I could only think of was my grandmother. My eyes started to close. It was as if the purity of the air, the bird songs, and the stillness of the surroundings had rocked me into a deep sleep.

I heard a beautiful voice singing the most beautiful song I have ever heard. Perplexed, I turned, and there was a woman who looked like my grandmother. Before I could open my mouth, she said- "My child, don't be afraid, I had wanted to visit you, but I am too old now and cannot hear when there is too much noise. But when I saw you today, and the fact that it was calm, I seized the opportunity. By the way, my child, tell me why is today so quiet?"

Me: Today is unusually quiet because there is a virus in the air, and we are under lockdown. People have been asked to stay at home so as not to spread it.

Grandma: My child, I am also happy to see you. When I used to sit here, I could see your grandfather's farm, my friend's house, the fence to our home, and the mosque we used to pray in. Now, I do not recognise anything except this baobab tree and the water spring below. What happened?

Me: This land is now what we call a mine. All these big metal pieces on wheels you see are cars, trucks, and trains. They are here to extract a mineral called bauxite, which is under the ground. This thing they are extracting is supposed to give the country a lot of money. Look, our grandfather's house is now the office of the mine manager. Your friend's house is part of the processing plant where a lot of noise comes from during the day. Our family's farm is one of the mines where the bauxite is extracted.

Grandma: I am confused, my child! This thing you call bauxite when you dig it, does it grow back? Before you answer, I'm thirsty, get me some water from the spring.

Me: No Grandma, it does not grow back. Everything is taken abroad. Sorry Grandma, you cannot drink the spring water; it is polluted. If you drink it, you will get sick. Here is a bottle of water. It is safer.

Grandma: The land is gone, and the spring water is no longer clean. Awa do no santideh - What has happened to this land, my child?

Me: Grandma, it is all happening in the name of modernisation. People have been made to believe that the money from the bauxite will give us a good life; that we'll have light, we just need to touch a button. And that for water, you just turn a tap. What's more, they said we will all drive machines and will not need to walk. However, in reality, as I speak to you, at home, I have no water, no light, no money, and not enough food. With this COVID-19 virus, I'm not sure my job in the factory is secure.

Grandma: So who owns the mine? When I tried to come out a while ago, I saw people who are much lighter than the Fulanis, like your grandfather. Some

were dark, but not like my cousins in Senegal. Are they real people, or am I having visions?

Me: The people you saw are from faraway lands called Asia, America, Australia, Europe, and the Middle East, and some are from other parts of Africa. They own the mines and have all come to work on them.

Grandma: And all these activities, what you call mining are being done for money?

Me: Yes.

Grandma: This is crazy, my child. Things cannot continue like this.

Me: I want things to change too; I am tired of the noise, the dust, the smoke from the machines. I am tired of hunger, the lack of money, and lack of control over our land and resources. I'm tired of seeing our spring being polluted, and I'm tired of not knowing if I will have a job tomorrow.

Grandma: My child remember this, when you are tired and want things to change, you can change things, but you cannot change what I see alone. First, it is essential to discuss these issues as a family and then as a community to come up with a solution.

Me: True.

Grandma: Hmm. Before our next meeting, I will leave you with an activity to do. Get a handful of seeds and plant them near your house. Count a handful of coins and put them in a room. See what happens! Let us meet after 30 days, and discuss what you see.

Me: Grandma, I will do what you said.

Grandma: See you in 31 days.

Me: Thank you, Grandma.

This is a fictional conversation written in loving memory of my grandmother who passed away on the 28th of April 1956 in Guinea. Guinea is located in West Africa. Guinea holds the world's largest deposit of bauxite and iron ore.

Frances Wall.
Clapping For Miners.

Think of something that needs mining? What did you think of – cars, metals, coal? Now think of a product or activity that does not need mining? What did you think of this time? Something organic - wood, an apple, cotton clothes – but think again - even if the product is organic and grown on a farm, someone still used metal tools for ploughing and for picking and transporting the products. Sewing machines and clothes factories all use lots in mined raw materials – even if you sew your own by hand, you still need a needle! Since the Stone Age, we have been taking advantage of the Earth's geological 'natural capital' to help us make and do things. It is an essential foundation to practically everything we do.

And now we need more – much more! – for two reasons. First there are more people on the planet who can afford electricity and consumer goods like cars, computers and washing machines. The second reason, and this is a really important one, is that the manufacture of the new clean technologies needed to combat climate change (think wind turbines, solar panels, electric cars) require more raw materials and a wider range of raw materials, especially metals, than ever before. The World Bank has coined the term 'Climate-Smart Mining' and estimates we need to mine as much copper in the next 25 years as we have mined in the last 5000 years.

Many times in school text books, I see sentences along the lines of, 'mining provides the raw materials for manufacturing but damages the environment'. So is mining something to be avoided if possible? Diagrams encouraging us to re-use and recycle to make a more circular economy often delete mining from their vision of the future – but as I said in the previous paragraph, mining is not an option, it is essential.

In many ways mining is like farming. This may seem an odd statement at first but there are about as many different kinds of mines as there are farms. Worldwide, farming ranges from large-scale automated and mechanised operations to family subsistence farming. Some farming is done well, in sympathy with the environment, and some causes lots of problems. Mines all over the world are the same. They range in scale, with at one end huge automated iron ore mines in northwest Australia, controlled from Perth thousands of miles away, with trucks as big as houses and their own railways and at the other, artisanal, family mines panning for gold or quarrying local building stone; with all shapes and sizes of quarries and underground mines in between. Some mines are run very well by companies who take great care about health and safety, the environment, sustainable development and financial governance, whilst at the other end of the scale, some are run by 'warlords' and gangsters.

The COVID-19 outbreak has demonstrated one way in which mining is not like farming. Most people would nominate farmers as essential key workers but I suspect that people were not clapping for 'miners' when they applauded on Thursday evenings. Actually, 'primary industry supplies' is on the UK Government list of critical workers under a section called 'utilities, communications and financial services' and this does include mining. But the urgency of mining for the well-being of people is less than food, even though raw materials are needed to make medical equipment like ventilators. I suspect, especially for us in the UK, mining is also a less urgent concern because most of our raw materials are mined abroad, 'out of sight - out of mind'.

Another way that mining differs from farming is that unlike food – which is eaten and gone and must be grown again, or coal and fuel oil that are burnt and lost to the atmosphere, metals such as copper, tin, lithium or neodymium are much more durable and can be used many times over, in re-used products and recycled materials. So, let's think a bit more about our metals and minerals and about where they are coming from and going to. Make sure they are mined well and then looked after.

Even if you
sew your own
by hand, you
still need a
needle!

Section 2.

For Earth.

"Neither the sea nor the forest so lends itself to the substantiation of the supernatural as does the mine"

Agricola, De Re Metallica

Henrietta Simson.
Transformative Wilderness.

Material Culture (1) is constructed out of porcelain paper clay and decorated with the traditional cobalt blue patterns that have travelled around the world in the form of the ceramic tile. From China to Turkey, to Europe and the Americas, these patterns have adapted, taking on local cultural identities – mutations driven by colonialism and global trade. The mountain form itself is also a kind of mutation, fusing Cornish waste heaps with the craggy outcrops of the painting of medieval Italy; its shape derived from both industrial and spiritual wilderness landscapes. It sits on a thin pewter plate, which produces a rippled reflection suggesting water and the flow of goods and ideas across the globe via rivers and seas.

The questions that inform this work focus on what these early paintings of landscape can mean today. As places that signify spiritual transformation, can they help to redefine the language that surrounds the mine, to shift the way we perceive it and its significances? Can an engagement with these early images help to integrate the industrial wilderness into our collective landscape imagination? An open-cast mine might be considered an 'eye-sore', a 'blot on the landscape', whether there is good or bad practice involved; its value measured (adversely) by the aesthetic constraints of traditional western landscape painting. The controversies that attach to mining often obscure the mine itself and we fail to value it and see its fundamental role in our lives. The work explores these challenges, drawing inspiration from attitudes to wilderness very far removed from our own.

The wilderness landscapes of medieval and early Renaissance painting were transformative spaces, associated as they were with visionary insight and change. Their features appeared to have mystical properties. Holy figures would atone within them, receiving divine wisdom in these remote and rocky sanctuaries. These were liminal spaces, unconnected to the rhythms of circadian time, places where the boundaries between dreaming and waking became blurred. Thought of as wounds and associated with healing, their caves and hollows nurtured the spiritual while sustaining the material dimensions of life. Like

portals, they linked the deepest, darkest recesses of the Earth to the heavenly sphere above. And the non-naturalistic style of these wilderness landscapes' depiction affirms this connection of body and spirit, for in spite of their spiritual associations their features are wholly suggestive of the material body. These are spaces very far removed from contemporary ideas of landscape. They are not characterised by a pleasing 'view', nor was their value measured in terms of the world of work – a place to go to restore the spirits in order to ready one for work again. Their scenery of rocky, barren outcrops, craggy mountains and caves doesn't function within the confines of capitalist production, with its work-life balance and the rational ordering of time and experience. Their cliffs and crevices – suggesting bony, abstinent bodies – appeal to a different sense of time and space. As depictions of landscape whereby the earth's 'body' is brought more in line with our own human bodies, they inspire empathy and a perception not based on any aestheticised view that acts as the backdrop to our leisure activities.

By moving away from this idea of landscape as an imaginary diversion from the reality of work and everyday life, we can begin to reshape our attitude to the mine. Industrial wilderness isn't considered a visually appealing place, and the traditional western landscape aesthetic obscures our ability to see any landscape that diverges from its code – even those places of industry needed to avert the existential crises of the 21st century. But we can reclassify the mine and bring it into the positive realm that other landscapes enjoy in our collective imagination. Seen through the medieval wilderness, the mine becomes a transformative space, where its material and spiritual impacts are brought to life. It becomes accepted into our frames of reference. As the blue tiles show, materials and visual forms are in continuous circulation. Cultural difference is just another point along a trade route which supports the flow of materials in our globalised systems. By engaging with the industrial wilderness – with the mine – as 'landscape', we can re-cast its image.

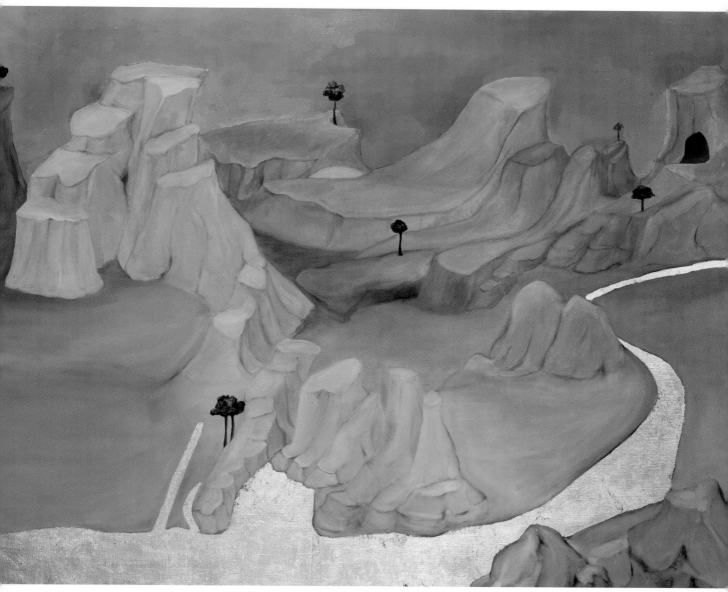

'After Thebaid' 120cm x 150cm oil and metal leaf and gesso on linen, 2020 Henrietta Simson.

O R E &

Bringing Natural Resource-Related Peacebuilding Down to Earth.
Bridget Storrie.

In 1937 the British travel writer Rebecca West visited the Stan Terg mine in Kosovo. At that time Stan Terg – now part of the huge Trepca mining complex - was owned and managed by a British mining company and West is enchanted by what she describes as the 'civilizing' influence of the mine on the local town. She's particularly taken with the Cornish-style mine houses, built with their windows confidently facing the road and with front gardens full of roses and sweet peas. After all, in a mining town, even in Kosovo (so the logic went) nobody need fear being attacked by their neighbours. Indeed, the Scottish general manager of the mine is proud that he employs both Albanians and Serbs and confident they will be able to work together. 'This country' he tells West 'is getting over its past nicely'.

Just over 80 years later, that little settlement above the mine is ruined, the houses cratered with bullet holes, the cinema where Albanians and Serbs used to watch movies together destroyed, the outdoor swimming pool high on the hill now an empty, overgrown carcass. I stand on the main street with an Albanian miner who used to sell milk and eggs to the Serb families here when he was a child, before the war. When the weather was wet, they would invite him in. We look at the empty-faced houses, the closed shops, the skeletal remains of the coffee shop. "I don't know what the hell happened", he says.

While the conflict between Serbia and Kosovo was not ostensibly associated with natural resources an

eight-day strike in the Stan Terg mine by Albanian miners protesting increasing discrimination against them by the Milosevic regime was one catalyst that led to the violent break-up of Yugoslavia and the war in Kosovo ten years later. Now Trepca is divided between Serb control of the mines and associated infrastructure in the Serb-majority municipalities to the north of the river Ibar and Albanian control of the mines in the south, including Stan Terg. But the complex wasn't designed to run like this and neither side is doing well as a result. Kosovo and Serbia both consider Trepca to be essential for their prosperity and agreeing the future of the complex is the most contentious, the most heated issue still to be settled between the two.

Meanwhile people live, work and raise their families amid the collapsing mine infrastructure and the ruins of the places Serbs and Albanians once enjoyed together. The material evidence, in other words, that the promises embedded in this orebody over time – that it would bring prosperity, peace and ethnic harmony – have all been broken. And yet, policy makers in the UN and the World Bank still argue that mining can bring peace to places recovering from violent conflict.

One of the problems with this is that it is a top-down prescription for peace, and top-down peace prescriptions don't necessarily achieve what they think they will, as Stan Terg shows. I argue that a 'bottom-up' approach is needed too; one that starts in the natural resource – the orebody - itself.

Because an orebody is not inert geology but woven into the ways in which people understand what a good life is, who that good life is for and how an orebody can help them achieve it. It is alive with the potential of the conflicting and converging promises we embed in it. It's like a rhizome, sending shoots above the surface, animated by what we want it to do for us. And like a rhizome (as Stan

P E A C E

Terg shows) it can run out of control.

As such an orebody is a place for thinking about what those promises are and where they are taking us. What sort of good life will they achieve? To quote Robert MacFarlane in his book Underland it is time to 'force ourselves to see more deeply' if we want to live better lives together on this planet.

Looking over the landscape around the Stan Terg mine it is clear that the the orebody plays a role in shaping people's lives here. It seems to add its geological heft to a project of territorialisation that divides people and keeps them separate and means certain people can't cross certain bridges – at least not easily – or attend certain churches or tend certain graves. There's an FCO travel advisory north of the River Ibar where the population is now mostly Serb, and the places associated with mining are the places that are the 'prickliest' – hung with flags, guarded by police, difficult to access.

There's a miners' monument in northern Kosovo, for instance, that commemorates the resistance of Albanian and Serb miners at Stan Terg to the German occupation during the Second World War, but it is difficult for Kosovan Albanians to visit now. I stood on the bridge that connects both sides of this divided city and asked my Albanian interpreter if we could go there together and he said we can. If we borrow his friend's car, take off the Kosovo number plates and if I do all the talking. 'If they know I'm Albanian', he adds, 'the Serbs will attack us with guns and knives, for sure.'

So, the orebody appears complicit in making the conflict here intractable. It is woven into the stories people tell about how things have turned out the way they have, who is to blame and what needs to be done about it. For the Albanian miners at the Stan Terg mine the orebody is associated with an idea of future prosperity that is closely aligned with

Kosovan independence and tied in with the history of the strike and the conflict that unfolded here. The story they tell is one of resistance, betrayal, sacrifice, struggle, loss, exile and their eventual return to the mine after the Serbs had left at the end of the war. That this orebody should benefit them now is the logical conclusion.

But another story emerged during the interviews I carried out with them. Alongside the conflictive narrative of blame and accusation another much quieter, more difficult, often transgressive story emerged of deep nostalgia for the pre-war Yugoslav past, of disillusion with fellow Albanians, of curiosity about former Serb colleagues and of uncertainty that the struggle of achieving independence has actually been worth it.

The orebody is therefore a place where narratives are both made and unmade, where identities are stabilized, and destabilized, and where something tentative exists alongside the intractable stories of the strike and the conflict that are told here. While geology seems to shore up the conflict dynamic that exists on one hand, it unsettles it on the other.

And that's something that should give a mediator pause. It's a curiosity. Like a geological anomaly it's a sign there may be something of value beneath the surface. There may be something worth metaphorically mining here.

Indeed, the further you travel down into the orebody the more evident this becomes. 750 metres below the troubled surface, there are no flags, no banners, no angry graffiti. Just the universal signaling system for the lift, the pipes and cables bringing power and air and a large poster advertising Rausch orange juice in the office where the miners get their instructions at the start of each shift. Apart from Me Fat Trepca Minetort (Good Luck Trepca Miners) sprayed in red on the wall, we could be under Cornwall.

And miners who worked here before the war, when this mine was mixed, describe how the ethnic identities that divided Albanians and Serbs on the surface fell away as they descended in the cage at the beginning of each shift, and how a deep friendliness and trust emerged between them underground. As one man said, "when you are inside the mine, ethnic groups don't exist". Instead they shared specific jokes, had certain conversations, treated each other in particular ways because they knew in an emergency it would be the person next to them who would help them, no matter who he was or where he was from. But this relationship didn't surface with them at the end of the shift. Instead as another miner said 'every conversation, every joke was left underground. When we got to the surface, we went our separate ways'.

Deeply embedded in this orebody therefore are the echoes of the conversations and the jokes with colleagues who are no longer allowed to be in this mine, and nostalgia for the days when they were. What is buried here is the reminder that alternatives exist to how things are on the surface. That there are other possible ways of being. And – perhaps - another natural resource-related future than the one that seems inevitable.

John Paul Lederach is one of the most prominent voices in peacebuilding. He says you should look for the strategic where of peace. For Lederach, these are places where people who are different from each other, who are 'not like minded' and 'not like situated' cross and come together – riverways, markets, schools, hospitals, highways - because it is these places, the relationships they hold and the ways in which they influence these relationships that have potential for building peace. That's partly because they engender interdependence. They throw people together in unexpected ways. They make us realize that while we may compete with other people, we rely on them too.

I think an orebody is a strategic where for peace. It has the capacity to hold together unusual relationships and it has the capacity to change them. It contains the seeds for positive social change. But an orebody is different from a riverway, a market, a school, a hospital, a highway. Unlike these places an orebody has its own geological story to tell about what is happening here.

Marcia Bjornerud has written a book called 'Timefulness. How Thinking like a Geologist can Help Save the World' In it she writes that we have forgotten that our personal and cultural stories have always been embedded in larger, longer – and still elapsing – Earth stories. The stories the Albanian miners tell me of the strike, the conflict and of their hopes for peace are embedded in a geological story that is perhaps about a different kind of conflict, a different kind of peace and that invites a different way of thinking about what it means to live well with our natural resources.

For Bjornerud, thinking like a geologist means thinking about social change in a way that goes beyond the life of the human. It means thinking past the urgency of the 'now'. It means thinking about the impact what we do now will have on the generations to come. What the orebody offers then is a place for reconsidering what it means to live well together not just now but far into the future – long after a mine has closed. I think this invites a longer, larger and more ecological kind of natural resource-based peacebuilding that doesn't assume that peace will come through the prosperity that comes through mining but asks what kind of long-term future people can collectively imagine. What geo-social peace can they envisage? And how can natural resources help achieve it?

The miners at Stan Terg have a sense that they are uniquely positioned to instigate change. 'The problems began here' one tells me 'So the solution can begin here too. The eyes of Kosovo are on us. The problem is we started to lose hope'.

But while natural resource-related peacebuilding might start with the miners at Stan Terg it doesn't end with them. After all, the problem of how we live well together with our natural resources is global as well as local. We all have a story to tell about what a good life is, who it is for and how natural resources – in our laptops, our smartphones, our antidepressants, our wind turbines - can help us to achieve it. We all metaphorically cross and come together in an orebody. It's a strategic where for all of us to consider what it means for people and the planet to live better lives together.

The orebody is ... a place where narratives are both made and unmade, where identities are stabilized, and destabilized, and where something tentative exists alongside the intractable stories of the strike and the conflict that are told here.

Dana Finch.
Visual Journey into the Dark.

Underground everything looked like a scene from a Tarkovsky film. We had entered a hole, poetically named 'the 725', in the side of the hill. A dark tunnel, overhung by jagged rocks, glistened in the light from many head lamps and smart phones. Silhouettes reflected in puddles. I was struck by the visual beauty of the inside of the mountain, with its striated edges drawing its own topography. We made our way in; underfoot varied from smooth and flat, to rocky, to muddy and slippery with vanishing footholds and sliding footsteps. Down and down we went, inside the hill. I was aware of the weight of the mountain above us - all its dense darkness. The air was damp and cool, but it felt clean, stripped of its usual meaning. The geologists among us stopped often to photograph bits of the rocky walls. They pointed out the seams of ore to me.

I had never been very interested in the lightless places beneath my feet. I am an artist, and in recent years my practice has been focused on painting. My first influence is light, but light carried and defined by the plants and other forms of the natural world. My visual language originated in a childhood spent in the English countryside, and the Moorish gardens of Spain. My parents took me every summer on long road trips through ochre plains and dramatic mountain ranges, where the stunning heat and dense shadows combined to make an indelible impression on me. Deserts and their vibrant contrasts have always called to me. But my understanding of the world was all above ground. So entering the world of mining was a challenging and unexpected experience. Very quickly I realized that the IMP@CT project would not be solely about the mechanical aspects of getting ore out of the ground. There was a dimension to the project that involved ethics; how it would affect people, places and the planet in a wider sense. It made connections to indigenous and local cultures and their ways of thinking about landscape, and place, as a site for cultural and individual existence. Our mine site for the project was based in Olovo, in Bosnia, and we had a field trip early on to visit the site and the community.

We emerged from the mine half way down the mountain, in a thick forest overlooking a valley, with a river winding far below through the steep banks of trees. All around the small clearing at the opening of the adit was plastic tape, looped over metal posts, and several danger signs, alerting us to the presence of landmines. The Bosnian war had been fought fiercely in these hills only twenty five years before. To me it seemed like yesterday – I was doing my art degree during the war and the horrific daily news had informed our work and our lives then. Signs of the war were everywhere, from the deserted houses, pock-marked with bullet holes and shell scars, to the abandoned workings of the old Soviet mine which had now been taken over by Mineco[1]. As an artist I have always been fascinated by ruins, by signs left of a lost life, a lost culture. Empty houses, abandoned buildings and scars on the land where buildings once stood all tell a fragmented tale and tie the present to the past. A ruin is a place of palimpsest[2], a place where one narrative is read through another, an overlaying of stories. Where the stories are within living memory it has an added poignancy. It leads into other adits, other mines, other hidden realities. Lost or fragmented culture is fertile soil for artists, trying to make sense of the forgotten, or the hidden, buried, underground and occult. I have a new awareness of our deep connection to this most ancient of endeavours. Our mineral life is tied to the Earth, and its many unfolded treasures. Our bodies are made of earth and the minerals it holds. Our psyches, our moods, our vigour all depend on the mineral balance in our blood. Our pigments are unearthed; our visual culture depends and has always depended on what we get from the ground. For me, painting is an investigation of memories, feelings and awareness of place. Themes return over and over. Light and land have recurred often, and now, the darkness, the underground, the meaning of mining and minerals, are emerging onto my canvases. I have sometimes described my painting process as a 'reverse excavation' and now that description makes more sense. Every painting is of Earth, and for Earth.

'Scenes from a goldmine' 100x100cm, oil on canvas, 2018 Dana Finch.

1. Mineco is the company who owns the mine at Olovo and is one of the partners on the IMP@CT project.
2. Craig Owens 'The allegorical impulse: Towards a theory of postmodernism'.
Art after modernism: (1984), New Museum of Contemporary Art.

Dominika Glogowski.
The Mine in Dialogue with the Environment through the Arts.

We mostly think of the arts in the context of extractivism either as a source to raise awareness for social and environmental injustice or as an act of revitalization, once the mine is doomed for closure. Yet the arts encourage us to pause, to listen, to look, to indulge in experience, to leave secure realms of specialization, to comprehend and to adventure the connection and reciprocal interconnection of all beings. The arts hence prompt us to engage with the space, the people, nature and the complex fabrics out of which the environment is created.

In extractive areas, notions of culture, nature, economy, wellbeing, but also toxicity and legacy are intertwined on a sensitive scale that has to be constantly balanced and negotiated. The mine is embedded into this glocal reality. It is interwoven with the intangible net, as it is physically with the landscape. Instead of an isolated industrial entity, the mine is shaping those fabrics like an organism. The fluidity of exchange and interdependency should be embraced as a chance for an active role of the extractive sector within its local setting to generate common identities and prosperity. For this to happen, spaces of dialogue, interaction and participation are needed. Whether in large-scale,

artisanal, switch on - switch off, or fly in - fly out mining approaches, the arts help to mitigate between "us" and "them" dichotomies. Bridging corporate headquarters with local populations, the arts can broaden the dialogue through embodied experience, activating a sense of place, belonging, community and coherent action.

Consequently, I argue with my think tank artEC/Oindustry for a new understanding of the mining industry. Apart from closed-off, seemingly alienated corporate structures, I reconceptualize the mine's (architectural) opening towards the environment. I thus challenge the incorporation of the arts into the early stages of a mine design as a creative tool for embracing wicked complexities between animate and inanimate stakeholders, inducing economic independency and foresight, empowerment and emotional wellbeing. Reflected beyond ongoing questions on transparency, empathy and meaning as new business pathways for industries, site-specific spaces for interconnectedness and participation could induce important steps towards dialogue, imagination and creation. In times of pandemics like Covid-19 such common, local and planetary future scenarios seem more than pressing.

Seirensho Art Museum, Inujima Japan 2011. Photo: Dominika Glogowski.

James Hankey.
dust >| dust

'At the same moment I had a pot and I did not have a pot'
This example demonstrates the whole law of the impermanence
of things. In particular it shows the human condition.
If this is so, I, the hermit Mila, will strive to meditate
without distraction.
The precious pot containing my riches,
Becomes my teacher in the very moment it breaks.
This lesson on the impermanence of things is a great marvel.'

A translation of a song by Milarepa, a Tibetan Buddhist poet from the 11th century, renowned for his ascetic lifestyle.

The title **dust >| dust** partly alludes to the phrase from the burial service in the Book of Common Prayer: 'we therefore commit this body to the ground, earth to earth, ashes to ashes, dust to dust; in sure and certain hope of the Resurrection to eternal life'.

Rare earth elements have an absolute permanence, they are created in very rare and hugely energetic celestial events, and are essentially indestructible. People are more fragile.

The title also references the energy consuming processes of crushing down the mined ore to dust-sized particles, essential for refining and purifying purposes.

Alongside these thoughts, I have gestured towards the pictographic symbol for a diode, >| . A diode is a manufactured component that is essential in every electrical device. Its function is to allow electrical current to only flow in one direction.

The combining of the cyclical nature of all things in the phrase 'dust to dust', with the one-directional symbol for a diode is an intentional juxtaposition indicating the paradoxical nature of mankind's 'progress' and the uneasiness in my thoughts towards the inevitability of mineral extraction.

For this commission I bought the latest iPhone 11+ for its super slow-motion camera capabilities to record the falling overburden into the disused mineshaft in United Downs. This area was once described as the 'richest square mile in the world' and helped fuel the Industrial Revolution. I then returned the phone within the fourteen-day returns period for a full refund, relaying to the Apple store attendant that I wasn't satisfied with its capabilities yet. This is true, but I always intended to return the phone. Perhaps it is I that needs to live a more ascetic lifestyle.

Above: Recording at United Downs, Cornwall, 2020 James Hankey.
Overleaf: Still from dust >| dust 2020 James Hankey.

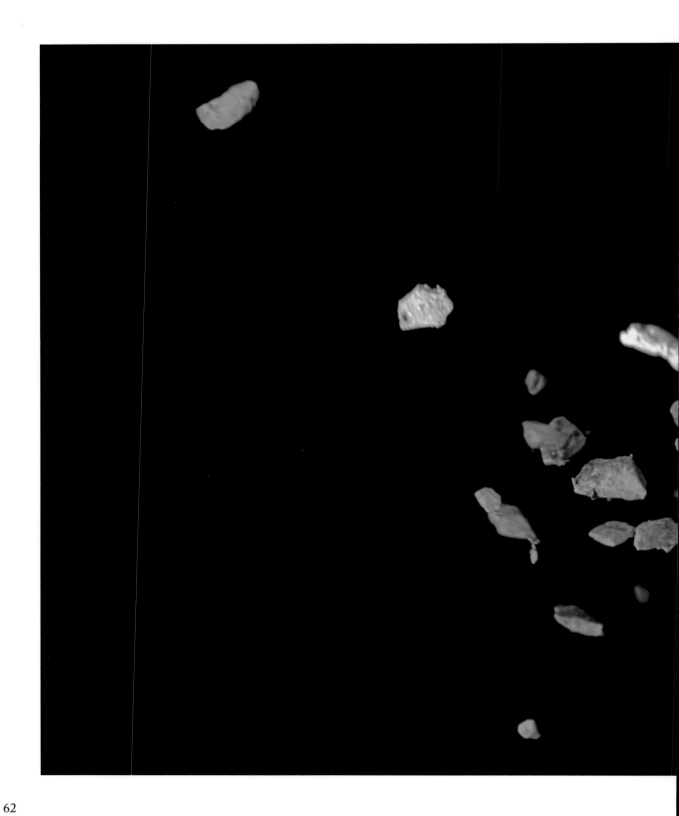

Kieran Ryan.
Who will listen?

*'We deserve a safe future: And we demand a safe future.
Is that really too much to ask?'*
Greta Thunberg; Global Climate Strike, New York, 20 September 2019

What kind of Earth will the young inherit? The theme resonates with A-level students interrogating the geography and geology curricula. At the core of these subjects lie the intricate links between the human and physical environment: global governance, the water and carbon cycle, and resource management included. I discuss with students the changes that humans are making to the planet and how the changes compare to natural changes in Earth systems throughout geologic time.

Students voice their frustration at inaction to change the relationship between humans and planet. There is change, but it is too little, too late and too slow. The challenges they will face in their lifetimes are mounting with each passing day. They see that their families (not of education age) are slow to adapt. How can they voice their concerns outside of their school and home? Who will listen? They are too young to vote, but old enough to object.

The extinction rebellion has appealed to very many students. Finally, they had a voice and an opportunity to be part of a local, national and global community to air their concerns and demand action. With this voice, they enacted change and their own college signed up to a county-wide green charter for schools. Students need more opportunities to speak out.

Participating in a 'Resourcing the Future' workshop, alongside the Of Earth, For Earth exhibition, gave students a unique opportunity to communicate directly with an international team of scientists and engineers working to support their sustainable future. The students talked to the exhibiting artists and fired questions about climate change, resources and manufacturing to the researchers who try to enact change at a government level. At the end of the workshop, the students were fuelled with excitement at the opportunity to convey their messages to the parts of society outside of the education system, where they feel the important messages about their future are not being heard.

The students reflected on their experience, when challenged with a similar brief to the artists. What is the conversation between self, material consumption and the Earth? How do young people envision a mine of the future? What is the meaning of a mine in the eyes of a student? Just three days before their school closed in response to the coronavirus threat, they returned their responses. The responses are those of young people who live in a historic mining district and appreciate the impacts of the withdrawal of mining on community.

Charlotte Luke.

What's mine is yours. 'It is important to recognise that by living our daily lives, we too are contributing to the pollution and habitat destruction we see in the news. We can no longer pass the blame onto corporations alone – we need to rethink our own consumption and take more of an active role in helping to make mining more sustainable.'

Maia Roberts.

Imagining a mine. 'We need to make sure there is balance between mining and the environment… in the future mining could even benefit the environment. Modern human life is sustained on our Earth due to it. A person stands taking a photo of the mine: The phone they are using also relies upon the mine to exist. Mines don't always destroy the environment around them. However the lack of plants [on a past mine site] also shows how changes should be made to make a less impactful mining form for today's society.'

Jacob Power.

'… mining is the bloodline of our society… no matter what happens in our world… we will unfortunately rely on mining, which would cause pollution.'

Bethan Jones.

'Mining is viewed in both the past and future, the brighter side being the future.'

Barney Martin.

A futuristic mine. 'Mining differs in the past and future when mining impact may be controlled remotely… As construction and excavation methods evolve, mining moves towards a safer, more sustainable future.'

Abbie Trubshaw.

A vision of the future. 'Complete blue sky shows there isn't pollution being produced by the mine. All tunnels are underground; only the entrance is present at the surface, reducing the destruction of habitats and wildlife and the amount of waste polluting the environment and landscape. The future mine can heal a scar on the land…'

These quotes relate to drawings made as part of this exercise.

Alison Cooke.
Heavy Lode.

Heavy Lode is made of clay dug from two Cornish tin mines. The clays were pressed into slabs and fired at a range of temperatures, some beyond melting point. The work takes visual inspiration from the granite outcrops of the Cornubian batholith and the mineral lodes that run through it. Using only pressure and heat to make the work, Alison's aim was to mirror geological forces that change materials from one state to another.

The installation was part of Tin Mine Clay, an art project in which Alison Cooke and Dominique Fuglistaller collected clay from Rosevale and Geevor tin mines and used it to make ceramics inspired by Cornwall's unique geology and tin mining heritage. At Rosevale, an ancient mine closed since 1910, deep within the adits, they were shown areas where different clays had been flushed through cracks in the rock and had collected naturally deep in the mine tunnels. At Geevor, now the Geevor Mine Museum, they dug old tailings ponds, the waste sludge, rich in iron and clay particles, left over from extracting cassiterite out of the pulverised rock.

The tin mine clays fired to strikingly different colours from deep terracottas to bright whites, their colours governed by the tiny particles of rock and minerals broken down over millions of years. Most were challenging to work with, being prone to collapsing, cracking and warping. Apart from sieving out stones and debris, the clays were used in their natural state.

The results of the Tin Mine Clay project were shown alongside photographs by archeologist Adam Sharpe, of the sites the clay was dug. The exhibition in the mill at Geevor Mine in 2019 coincided with the centennial of the sinking of Geevor's Victory shaft and the Levant mining disaster.

With special thanks to the Rosevale Historical Mining Society and Geevor Mine Museum.

'Heavy Lode' installation view, 2019 Alison Cooke.

Karin & Chris Easton.
Prosperity and the Perils of Romanticising Mining History.

Perranporth, on the north coast of Cornwall in the south west of England, has grown to become the largest village in the parish of Perranzabuloe due to the influence of tourism since the late 1800s.

Although there is evidence of widespread mining in the area prior to this, substantial mining in Perranporth ceased early as a result of a legal dispute between the two main mines over the boundary for underground mining rights. Local entrepreneurs, in an effort to revive the area and attract holiday makers, destroyed much of the evidence of mining, with its engine houses and chimneys. During the Second World War, Cligga Head was worked for tungsten and tin, but further mining evidence was destroyed as possible landmarks for invading German forces. So, during the twentieth century the area became known less for mining and increasingly as a desirable holiday destination.

Despite the early cessation of mining and loss of its infrastructure, local place names still provide many clues to Perranporth's mining past. Names such as Wheal Leisure and St George, the two previously-mentioned mines involved in the legal dispute, are now the names of roads, housing, a hotel and a carpark, the carpark being on part of the site of Wheal Leisure. Author Winston Graham, who lived in Perranporth, used the name Wheal Leisure for a mine in his popular Poldark stories. Local pride in Perranporth's mining past has resulted in more recent developments using local mining nomenclature such as a new social housing development called Wheal Catherine Close.

Individual house names can also tell us about Perranporth's mining past when the successful returning miners called their houses after the names where they had been working abroad, names such as Massiac (in France), Bel Fiori (in Italy), Kimberley and Virginia House, all evidence of the Cornish diaspora. Perhaps this is some of the reason for the romanticising of Perranporth's mining past as it shows financial success as well as a sense of adventure and the exotic. Oral history from a local family tells of a family member returning with a parrot, and in its collection Perranzabuloe Museum has an elephant's tooth, found in an attic.

This presents us with the paradox of romanticising the mining past without realising that there is still evidence of mining in the cliffs. What visitors and even some residents take to be sea caves are almost all mine workings. This resulted in the unfortunate death of an eight-year-old girl when her father thought they were going to explore a natural cave. Beach users, despite warning signs, seem to remain unaware of the instability of the cliffs resulting from mining, not just natural erosion.

Perranzabuloe Museum and the University of Exeter set up a working group to investigate how much beach users understood the mining heritage of Perranporth and whether the cliffs were natural or "man-made". The results of the survey showed that most people had no realisation that the "natural" arches and caves were the result of mining. Arising from this collaboration was the Heritage on the Beach project to inform beach users about Perranporth's mining heritage in the cliffscape by taking the museum on to the beach.

While there has been no mining in the parish since the early twentieth century, with the exception of short-term mining at Cligga Head, the impact of mining is still felt within Perranporth both bad and good, the bad being the dangerous caves and erosion of the cliffs and the good being the romantic view of mining and its benefit to tourism and to local identity.

Droskyn, Cornwall Photo: Kathryn Moore.

Nic Barcza.
A Life of Mining.

My first exposure to the mining industry was at the age of around 5 years when I became aware that my father was a mining engineer on a gold mine near Johannesburg in South Africa. We lived in the mine housing estate with a large garden and lots of exciting things to do there.

My schooling was at a Boys' College founded by two Cornish entrepreneurs Albert Collins from the village of Stithian and William Mount Stevens, who both emigrated to South Africa during the years of rapid development of the Witwatersrand gold mining industry. They established a successful building business and bequeathed their estates for the establishment of a Methodist school, St Stithians College. The school badge is based on the coat of arms of the Duke of Cornwall and has the same motto, 'One and All'. I was a foundation scholar there and our class still meets up at the school about once a year.

My exposure to mining continued when we spent school holidays at Pilgrim's Rest, the historical gold mining village near to the famous Kruger Park game reserve. In fact a project during my final year at school was on the 'Valley of Gold', based on the book by A P Cartwright. Pilgrim's Rest combined the beauty of nature in the region with gold mining and I recall the excitement of panning for gold in the Blyde (Dutch for happy) River.

This exposure to mining encouraged me to study Geology at senior school and I soon realised how much this opens one's eyes to one's surroundings. I continued with Geology at the University of the Witwatersrand as part of a metallurgical engineering degree that I felt complemented mining and it resulted in a wider appreciation of resources. I was privileged to go on an excursion to the Bushveld Complex led by the world famous Paleoanthropologist, Professor Phillip Tobias and during that trip saw for the first time a chromite deposit not that far from Pilgrim's Rest. This event lead to one of my main areas of involvement in research and development for which I am eternally grateful.

The sampling of chromites from a number of the deposits at the mining operations in South Africa provided the basis to work on the conversion of ore into a chromium iron alloy that is used to make stainless steel. This gave me a good understanding of how mining and processing of chromite provides us with the stainless steel that is used in so many ways in our everyday lives. It has so many benefits, is fully recyclable and can be used in numerous applications: construction, domestic and household, medicine, agriculture, the chemical industries, and artworks such as statues and memorials. The name chrome comes from the Greek word khrōma 'colour' since chromium compounds have a range of bright colours some of which have been used in artwork and paints over many years. The experience of working with chromite mining, downstream beneficiation and uses, and appreciating the benefits that various stainless steel products offers humankind, have been truly rewarding and it's gratifying to see how the current generation is continuing with this imperative.

Nic Clift.
A Natural Land.

Many of those who challenge the acceptability of mining have firm a priori views about Nature, mining, and the moral implications of human involvement of any kind. "Nature" is almost always imagined as something between pristine Amazon rainforest, and the idealised orderly green fields and pastures, filled with farm animals and surrounded by diverse ecosystems in hedgerows and copses, so typical of much of (for example) Europe's agriculture. Molten rocks and ejecta from volcanic activity, and the poisonous and sulphurous fumes that often accompany them, are "natural" and thus a thing of great beauty worthy of awe and respect. On the other hand, slag from a furnace, or the treatment of sulphurous fumes in industrial processes, which may be chemically very similar, result from human activity, and as a consequence are viewed negatively. A group of environmentalists I presented to in Bathurst some years ago, liked the idea of mining asteroids, but (angrily!) considered leaving residual gangue in the asteroid belt, or allowing it to fall into the sun, to be carrying out unacceptable pollution of space.

My anecdote considers the impact of mining on indigenous people, previously trapped in meagre subsistence farming and conflictual seasonal migration, where hungry villagers have a devastating impact on the surrounding flora and fauna.

Guinea, in West Africa, is home to an extensive bauxite mining industry. The richest bauxite is found in huge surface deposits, on the Sangaredi Plateau, that are many tens of metres thick. The bauxite is of high purity, in situ, causing very large areas of impermeable ground. Some vegetation establishes a toe-hold in streams and rivulets that form due to surface drainage. Otherwise there is only a thin coverage of tall spindly grass that grows rapidly after seasonal rain and then quickly dies and dries to leave bare ground that is almost a lunar landscape, and does not provide sufficient food for cattle during the dry season. As areas of these vast deposits of bauxite are exhausted, the mining company – CBG (Compagnie des Bauxites de Guinée) breaks up the ground with rippers; spreads around 30 cm of topsoil; plants a range of seeds typical of the flora of the region; and plants anacardia (cashew) trees. All of these grow profusely, creating a lush terrain capable of supporting far greater quantities of livestock – particularly cattle – than the land could ever manage in its "natural" state. By 2007, over a million anacardia trees had been planted.

It is impossible that human beings should live on this planet and leave no footprints. Nor is it a reasonable nor rational expectation that we should. My view of sustainability corresponds with that use of the term that has its roots in the 1972 United Nations Conference on the Human Environment in Stockholm, which focused on what was needed "to maintain the earth as a place suitable for human life not only now but for future generations" (Ward and Dubos, 1972). In the case of this anecdote, mining has definitely left the land in a changed state. The mineral deposit has been exhausted locally, bringing once-off income to the government, and is no longer available for future generations. However, arguably, greening of the exhausted areas of the mine brings greatly enhanced enduring new value to the local people. In this case the impact of CBG's mining activity is thus both positive and sustainable.

'Material Culture 1' 40x70x65cm, porcelain paper clay, ink, pewter, 2020 Henrietta Simson.

Clay detail from 'Material Culture 2', 130x80x30cm, clay, paper clay, photographic acrylic print, 2020 Henrietta Simson.

Djibo Seydou & Naomi Binta Stansly.
Niger - The Road to Sustainable Mining.

Niger is a landlocked country at the crossroads of the Sahel and the Sahara, rich in mineral resources: uranium, coal, gold, tin, lead, zinc, and iron. Uranium, gold, and coal mining are the major mining activities for the country. Niger is the fifth-largest uranium producer in the world. Uranium mining activities started in 1971 with the exploitation of the Arlit uranium deposit by the Société des Mines de l'Aïr (SOMAÏR). In 1978, another uranium mining activity was launched in Akouta by the Compagnie Minière d'Akouta (COMINAK). The Anou-Araren coal deposit was put into operation in 1981 by the Nigerien Coal Company of Anou-Araren (SONICHAR) to supply these two companies with electricity. The artisanal exploitation of gold began in 1984 in Niger, following the discovery of gold at Tchalkam in the Nigerien Liptako. Industrial gold mining started in 2004 with the development of the deposit of Samira Hill by the Société des Mines du Liptako (SML) near Tchalkam, where gold has been discovered for the first time in Niger.

Uranium mining wastes have been stored for more than 50 years in the open air near to local communities. Several people living in towns created by mining activities, including Arlit and Akokan, attribute their health problems to the waste from uranium mining. The exploitation of industrial gold has led to the creation of several open pits and mountains of mining wastes: tailings and waste rock next to one of the main tributaries of the Niger River, which gives rise to fears that water resources will be contaminated.

In parallel with the industrial exploitation of gold, the artisanal exploitation of gold or gold panning has developed in the Liptako, but, also in the Sahara since 2014, this activity is carried out in an anarchic way without respecting health and safety rules and often without prior authorization. Unregulated and lawless artisanal and small-scale gold exploitation causes enormous damage to people and the environment.

Despite the wealth of mineral resources that the country possesses, weak governance of mining means that the people of Niger and local communities do not benefit from the exploitation of these resources. The exploitation of uranium, gold, and coal has not contributed significantly to the socio-economic development of Niger. In the majority of the country, mining regions lack roads, clean drinking water, electricity, health centers, and schools. In 2007, Tuaregs from Northern Niger rebelled against the Nigerien government, asking for a fair share of uranium revenues.

Niger has experienced significant social tensions due to the exploitation of mineral resources. Mineral resource exploitation should be done responsibly and sustainably to bring various positive benefits to host communities, particularly access to education, employment, infrastructure, and health care and avoiding negative social issues. A secure link needs to be established between government, local communities, and mining companies for information sharing, and decision making. The vision for better governance of mining in Niger must be based on a mechanism of regular consultation that seeks to optimize the economic benefits of mining for local communities and society as a whole, while minimizing negative environmental and social impacts.

In the majority
of the country,
mining regions
lack roads, clean
drinking water,
electricity, health
centers, and
schools.

Uranium, some believed, was a white rock under the earth...

Richard Martin.
Power, Poison and Hope in the Border Country.

Indigenous people are often thought to be anti-mining. This arises from the very commonly-held view that Indigenous people have a special relationship with nature, and may in fact hold the key to creating a more sustainable way of being in the world. This idea finds expression in Indigenous rights advocacy, as well as popular representations of Indigenous people on tv shows and films like Avatar, in books, and even on tea-towels reminding people that they cannot eat money. This is harmless in most instances and may even be helpful in the context of particular Indigenous people's struggles against particular developments, including unwanted mines. But it tells us very little about how Indigenous people actually think about mining.

In 2007, at the very beginning of fieldwork for my PhD about the remote Gulf Country of northern Australia, I was employed by an Aboriginal Land Council to undertake a cultural heritage survey about a prospective uranium mine development. It was only my second trip to the Gulf Country, and I was as green as a new shoot of grass. Driving into Burketown for the first time, I almost got lost on the featureless plain – empty of landmarks apart from the odd cruciform shapes of Parkinsonia and mesquite trees in the yellowed grasslands.

My first job in Burketown was to find Danny Wollogorang, a senior Garawa man from the Northern Territory who had travelled across the Gulf

with another man to meet me. The other man, his cousin (or 'cousin-brother'), would come to adopt me as a brother before he died (and so he is here unnamed). These men were to accompany me back to the border, where the grasslands of northwest Queensland give way to the forested hills and gorges of the Territory. A mining company was proposing to drill a series of holes amongst the gorges in its search for uranium, and Queensland's cultural heritage legislation required the company to consult with Aboriginal people to prevent damage to their sacred places, so Garawa and Gangalidda people had organised a team of senior men to survey the area with me, an anthropologist, as well as a representative from the mine.

Danny, and the man I called brother, knew the country between Burketown and the border, and as we drove west they explained it to me. There are Dreamings in the landscape, they said: kangaroo, 'left-hand' wallaby, black-headed snake, bushfire. What does that mean? I asked, as I imagined a bushfire bursting into life out of the ground with black-headed snakes fleeing before it. I can't recall the answer I received; I came to understand that the Dreaming just was, and is, like the land and the sky.

At an isolated homestead named Hells Gate near the border, we were joined by other Garawa men who had travelled from the Territory and some Ganggalida men from Burketown who arrived before us. We drove through the bush to a camp-site away from the homestead, amongst the hills. They showed me ancestors imprinted on the rockface in ochre and spit, and we spoke more about the Dreaming. We also spoke a little about mining, and I explained the company's proposal to drill. Later, after a dinner of barramundi roasted on coals, I fell asleep in my swag while the old men from the Territory swapped stories with the younger blokes from Queensland. Occasionally they broke into song.

In the morning, a helicopter arrived with a mining engineer and we flew around the tenement where they were looking for uranium. I wrote down what I was told and passed it on to the Land Council. The places where they wished to drill were replete with many sacred sites which the company, in due course, promised to avoid.

Over the following months, as I began my fieldwork in earnest, I heard much talk about this mine, and the uranium being sought. Uranium, some believed, was a white rock under the earth that could be 'sung' and used for sorcery. Without this song, it was just white rock, but people had heard stories and they knew that it was dangerous. For some, this meant the company should not mine. But for most, the danger

posed by this rock had to be balanced against the potential benefits of mining. 'We need this mine', a young man whom I called son told me: he told me they would build a hospital with the money from the mine and buy a dialysis machine to bring the old people back from the city and look after them in the bush where they belong. 'Aunty, in Darwin, we're going to bring her home'. 'We are not going to waste this opportunity'. Older people were generally more circumspect about the mine, but they worried about the young people and were mindful of their views.

Years later, there is no mine, and the border country remains much as it was when I saw it in 2007. The young man who dreamt of building a hospital tragically died too young, around my age. Perhaps his life could have been saved by a hospital, or just by a job at the mine. But the uranium price is too low, so the deposit remains under the ground, buried amongst the Dreamings, waiting for the song to spark it to life.

Oliver Raymond-Barker & Caitlin De Silvey.
Natural Alchemy / Elemental Analysis.

Geologist J. B. Hill visited the Carnon Valley in 1902 during a survey of geological deposits and associated industries in the region of Falmouth and Truro. Tin and copper production had dropped off precipitously and the industry was in decline, but the sides of the valley were still pocked and punctured by old excavations, and by adits that drained unwanted water from the network of deep mine shafts. The most extensive drainage system, the Great County Adit, discharged millions of gallons of water a day into the Carnon River through its outflow portal, the terminus of over 65 kilometres of snaking tunnels.

At Bissoe, just downstream of the portal, Hill came across a cluster of auxiliary industries that had developed to take advantage of the mining industry's excess. Two arsenic factories operated near the Bissoe bridge, refining the crude arsenic from the mines for use in agriculture, pharmaceuticals and manufacturing. Hill wrote, "Ochre works exist in the same locality, which at the time of our survey in 1902 were said to have been in continuous operation for over 40 years. The oxide of iron carried in suspension, in the waters that discharge from the main adit into the Carnon stream, is caught in small pools, after which it is pulverized and washed. It is then dried into a fine brown ochre, in winter by fires, and in summer by the sun. The works are small, and there is no continuous market for the product." Occasional buyers included the fishing industry, which combined the ochre with oil and coated sails with the mixture to protect them from seawater.

Mineral-rich water still flows from the Great County Adit's exit portal near Twelveheads, carrying subterranean residues from the exposed surfaces of the old mines into the daylight world of the hard-used creek. The sediment that accumulates on its banks is a deep rusty red. A little over a century after Hill's visit, Oliver scrambled down the bramble-choked banks of the river to scoop up a sample of sludge. He brought it back to his studio, where he spread it on glass plates to dry in the sun. He then ground the grain into a finer dust with a mortar and pestle, and mulled the powder in linseed oil. The resulting ochre pigment colours the images and the words you are reading. Inadvertently, Oliver replicated almost exactly the process that Bissoe's ochre works had used to transform mining waste into practical matter.

The pigment carries the traces of its origins. An elemental analysis shows high concentrations of arsenic and other heavy metals. Boats with ochre-red sails still ply the River Fal downstream, although the traditional fisheries are under threat from the accumulated effects of post-industrial and contemporary contamination. The exchange continues - earth elements suspended and then seized for our purposes, spinning out unintended effects, spilling into a creek, onto a page.

Text commissioned for the publication Natural Alchemy, 2014. The project was a collaboration between Dr. Chris Bryan of the University of Exeter's Environment and Sustainability Institute (ESI) and artist Oliver Raymond-Barker. All text copyright the author.

'Natural Alchemy, Wheal Jane process facility', 3.5 x 5.5 inches,
photo etching on paper printed with ochre from the Carnon river,
2015 Oliver Raymond-Barker.

Section 3.

Un

Earth.

Gill Juleff.
Colour Mines.

When we talk about the legacy or heritage of mining we conjure images from a highly visible past industrial era. These may be of either the nostalgic and heroic silhouettes of great buildings and engineered structures from an era of new wealth, or the painful scars of poverty and wounded landscapes. As an archaeologist of mining and metallurgy, when I think of the heritage of mining I cast my mind much further back in time, many thousands of years, to when people first engaged with those places in the landscape where they could see and gather dense, brightly-coloured rocks, and humanity began a deep and enduring relationship with the mineral world. At first, these places were sources of trinkets, idiosyncrasies of nature that could be carried home to camp, to be bartered and exchanged and in time to become the starting materials for beads or pigments. The blues and greens of copper minerals, or the reds, oranges and earth ochres of iron, the blue of lapis and the glister of gold. The value of these minerals lay not only in their integral beauty but also in their origin. The places in the landscape where they could be found might be marked by distinct features of topography, the lay of the land, or changes in vegetation, perhaps attracting wild animals, or repelling them. Whatever their characteristics, these places would become important, possibly sacred, or secret to only a few. We could then consider these first mineral places as 'colour mines'. Over millennia, familiarity with the behaviour of these minerals led to their transformation, in the presence of fire, into new materials. The great, slow-motion, technological leap forward into metallurgy changed forever the social world of humans.

Over the passage of time people have exercised choice in where and how they engage with their environment. Where to farm, where to settle and create villages, towns and cities. There are few environments that have constrained humankind. Even though you can only be a fisherman in the sea, you have a multitude of choices on how and where you access the sea. The landscapes we see around us today are shaped by these choices. Perhaps only the mineral landscape imposes limitations on our choices. We can only mine where the minerals occur and miners can only live where minerals can be mined. And across the world these places are sparsely distributed. We can list where copper, or tin, or gold, and the rare minerals that often accompany them, can be found. This concept fascinates me. It means that miners must return to the same places generation after generation, millennia after millennia, to search for the same minerals. We now have new tools to search with and we can probe deeper into the earth but in doing so we are very likely to encounter along the way evidence of these ancestral miners. They may be from the first chalcolithic era, the Bronze Age, Roman or Medieval times. The marks they have left on rock faces and the tools they abandoned tell us about how they worked and who they were. How, like us, they sought out the richest minerals and took away the least overburden. This scant and precious evidence can only be found at the mine and it is vulnerable to the work of all the miners who come later and who have the option to respect or destroy it.

There is hardly a culture in the world that doesn't have a name for these 'old men's workings' that chart the lives of grandfathers, great-grandfathers and generations of ancestors. Visiting a Mao-era mercury mining complex in China, restored as a theme park attraction, it is possible to find just meters away from a vast mining cavern now used as a brightly-lit wedding venue, small pockets in the rock face created by fire-setting, one the oldest known mining techniques. On close inspection, the same can be found in mines across the old world from Iran and Anatolian Turkey, the Balkans the Alps and Iberia, to the Atlantic fringes of Britain and Ireland.

Some decades ago I would have despaired that any of this valuable legacy could survive the brutalist plunder known as modern mining. But our planet is changing and we are revaluing our relationship with our environment. With innovation and sensitivity we have opportunities to reconnect with our mining forefathers of a deeper past, and hear their stories.

Parys Mountain, Anglesey Photo: Kathryn Moore.

Heidi Flaxman.
Carn Brea: An Alternative Landscape.

As an artist I am influenced by landscape and its intangibility, though it exists physically around us. Only a small element of landscape can be experienced by an individual at any one time, it is the culmination of a nexus of features which is perceived from one perspective, that of the viewer. My creative practice investigates the intangible through tangible artefacts, which raises questions regarding the role of art and contemporary craft in assimilation of environmental awareness. This awareness may include the role of landscape and mining, and incite dialogues relating to the sustainable futures of extractive practice and landscape conservation.

The processes of the Anthropocene scar landscapes, leaving visual disruptions that may be irreversible. People may feel alienated from the mining landscape through ecological change, accidents and pollution, or through the technical and scientific processes of extractive industry where understanding is lost.

Three pieces were created for Of Earth, For Earth. They offer an alternative perspective of a landscape, Carn Brea, Cornwall. This area was chosen due to proximity of the final IMP@CT conference and the rich mining history of the area. The pieces are intended to show the landscape and the mineral wealth within.

Traditional mining practice - The first piece represents a traditional concept of mining where it was imperative to continue production to maintain financial viability. The surface of the landscape is saturated in mineral deposits reflecting abundance below the surface. Extraction intensity is greatest here. The landscape is worn.

The static landscape – The second piece expresses a visual equilibrium where the land, its contents and the ecology are in balance once extraction has stopped. Mining has reduced the mineral richness of the Earth, though unexplored areas remain.

Sustainable mining - The third piece in the series visually presents the concept of IMP@CT's Switch On Switch Off (SOSO) sustainable mining paradigm. The land is not depleted, nor over processed. The minerals are there within the earth, extraction is specific. This method is sensitive to availability, which is suggested through lesser saturation of pattern. Allowing higher grade materials to be removed from the earth when required and to also allow the landscape to recover.

This series of work has been created with the intention of developing connection to the landscape, through deeper understanding of the physical geography and geology. Visual presentation of the unseen and intangible through craft aims to enhance these connections through the tangible artefact; invoking feelings of 'care' towards the area which is necessary for a sustainable future of these heritage sites.

Facing page:'50°13'18.5"N 5°14'48.3"W Carn Brea, Cornwall. Fine grained vein in coarse granite' Copper infused PLA filament, cotton warp, wool weft, 2020 Heidi Flaxman.

Anshul Paneri.
Clean Energy for Mines.

Being a renewable energy engineer, I look for industries and societies where renewables can bring a change for the community and environment. I have also been looking into sectors which are harming the climate. Some of the biggest greenhouse gas emitters are mining companies. Mining industries consume a substantial share of global electricity production. Until the last decade, most of the electricity demand was provided by fossil fuels from grid-supplied or on-site power generation using diesel or other fuels. This is changing now, mining industries are now focusing on sustainability, and clean electricity generation is one major part of it.

Renewable energy has been playing an essential role in providing clean energy to the mines. Almost all mining processes can be electrified, which can save millions of litres of fuel and reduce tons of greenhouse gases emissions. Earlier renewables were more expensive than fossil fuels, but this has changed. Now the mining industry is recognizing that with the adoption of renewables, they can work towards sustainable mining. Large scale mines in Australia, Chile and South Africa have already integrated their power generation with renewables and are showing benefits of switching to renewable power. In remote areas where there is no cheap grid power supply, the mines are entirely dependent on diesel fuel for power generation. In such conditions, renewable power generation is showing great potential in providing cheap and clean energy to the mines. Renewables can also help the social communities located near the mines. Once the mining is over, the renewable power plants can be handed over to the local community for their power supply needs. Some renewables such as biomass generators can help in developing a small scale circular economy for the communities based in remote areas.

Even though renewables can benefit mining and help in reaching the sustainable environment for the industry, there are still many challenges. Fossil fuels are heavily subsidised in many countries for industrial purpose in comparison to the renewables. This creates a void between renewables and fossil fuels where, even though the renewables are cost-effective in comparison to diesel under normal circumstances, sometimes diesel power generation is still preferred because of subsidies. A lack of new policies and regulations to promote renewable technologies is another reason where renewables are facing challenges. The governments have to take initiatives by regularly updating the renewable policies, and providing subsidies to diesel only where it is essential.

For the first time in history, the crude oil price in some parts of the world is negative and oil suppliers are effectively paying consumers to take oil out of storage. There has been overproduction of oil in the face of rapid economic decline during the Covid-19 pandemic. The raw materials production sectors (fossil fuel and mining) are characterised by demand and price volatility. In a framework of exceptional times and exceptionally low oil prices, it is difficult to imagine how any other energy source can compete economically with fossil fuels. But could volatility in the production sectors be offset by the use of clean energy? And economic considerations are not the only factors at play. What is the cost of limiting climate change? And how is this factored into environmental impact assessments for mining?

Cassia Lynn Johnson.
Mining Inequality through Time and Space.

My love of nature is intrinsic to who I am. I explored this aspect of myself academically when I pursued a degree in geological sciences. The study of the earth, with its vast systems and spectrums, has continued to fascinate me throughout my career as a geologist. As someone who is also an environmentalist, I have had an inner struggle working in the mining sector. I attribute my inner struggle to how I might be, and have been, perceived from an outsider's perspective. How could someone who feels connected to nature be employed for the purpose of its exploitation? As I have progressed through my career, and have grown to understand the nuance, it has become clearer to me that it is essential that environmentalists work on the inside of the mining industry to ensure responsible, sustainable practice. I have found a deep interest in the role mining has in society, and it has inspired me to do PhD research of the potential for the industry to contribute to sustainable development.

Human development is connected with resource extraction. From the Stone Age, with the use of stones for tools, to the modern Digital Age, and the use of battery metals for technologies, what is mined reflects human progress. Not only is mining linked to human advancement, but also the development of nations. The 19th century gold rushes in Canada, USA, and Australia, for example, provided the foundations of growth for 'first world' nations. The artisanal and small-scale mining (ASM) of the gold rushes, which was legal and encouraged, was fundamental in building infrastructure such as rail and road networks, as well as banking systems. The ASM in many developed nations also contributed to a wealth of geological and mining knowledge, allowing for the development of large-scale mining operations and legal frameworks to regulate and govern the industry. Although the rushes leveraged development, it was sometimes at the cost of the environment and existing social structures, such as deforestation and pollution, or mistreatment of Indigenous communities echoing colonialism.

The ASM gold rushes of the 19th century continue today in some developed nations, such as Canada, but predominantly in developing nations, such as Bolivia, Ghana and the Philippines. In many developing countries, ASM is a major, and sometimes the only, source of income in rural areas. The modern gold rushes are not the romanticized tales of the 19th century. Present ASM operations in developing nations are often discouraged by governments and sometimes even illegal, instead of vigorously promoted as they were in the 19th century. They are often in direct competition with large scale mining operations, either neighbouring mines or mines elsewhere, operated by global companies based in developed nations. And they are under high levels of public scrutiny for the environmental and social problems that are often associated with the industry. The rights to the earth's riches, and how the riches are unearthed, is not simple.

Despite the complications, ASM is an opportunity to better the lives of millions of people globally. Platitudes are issued that purport to mitigate the environmental, social and economic challenges of ASM with the goal. But tangible working solutions that are tailored to context are required to realise the goal of sustainable industrial development. With my research, I hope to add to this industrial metamorphosis: to contribute knowledge on how to continually develop ASM into a sustainable industry. The metamorphosis of ASM from an illicit industry to a sustainable builder of nations will happen within the global context of a spectrum of mining operations, where resource extraction operates unequally across time and (developmental) space.

Heather Wilson.
Mining as an Affront to the Planet.

A few years ago I came across an article about a weaver and subsistence farmer, Maxima Acuna, who lives and farms in the mountains in northern Peru. I was shocked to learn of the massive open pit gold mines next to her home, and that she and her family were being seriously persecuted for refusing to allow their own land to be destroyed for gold mining. This inspired me to learn more about the extraction and mining industry, as well as about the environment in general, and this has become the central focus in the artworks that I make.

I consider/have considered mining as an affront to the welfare of the whole planet that we depend upon, from mountain removal, to deforestation, and to seabed destruction. I want to see the scale of mining and its impacts tackled with absolute urgency. I abhor the pollution caused by mining, the resultant tailings and also landslides. The impact of mining on the safety and security of people and communities concerns me greatly.

Making sure that we attain a balance in how we utilise the earth's resources is crucial. I am interested in the impact of mining and extraction on communities and indigenous people as well as the impact on flora and fauna. Any serious attempts to ensure that negative, destructive practices are not just kept to a minimum, but stopped, is essential to preserving the environment for future generations, and to making Earth a habitable home, conducive to the coexistence of humans, the natural world and other species who share their planet with us. The destructive mining environment is tied into the disassociation of people from their own environment, where fauna and flora are of little concern and profit is considered important above all else, even human life.

My inspiration comes from the land, and my work is a combination of memory, experience and research, into land use and also land abuse. I embrace the idea of dialogue about mining practices. The careful management and regulation of any essential mining in the 21st century is crucial to protecting and preserving the ecosystem of the Earth.

'Maxima Acuna' mixed media on paper, 2020 Heather Wilson.

Allie Mitchell, Joel Gill and Nic Bilham.
Sustainability-Driven Geoscientists Making a Difference.

Ever since its emergence as a science over 200 years ago, geology has played vital roles in economic and societal development, from sourcing the raw materials that fuelled the industrial revolution to improving our understanding and mitigation of natural hazards. But the impacts of humankind on our planet have intensified, and we are becoming increasingly aware of those impacts, including dangerous climate change (with fossil fuel use a major contributor), as well as the negative social and environmental consequences that have all too often arisen from mining. The UN Sustainable Development Goals (SDGs) represent a global call to action for all, including the geoscience community. Crucially, however rapidly we progress towards a 'circular economy', we will need an ever-growing range of raw materials from the ground, for low-carbon energy and smart digital technologies, and to secure equitable access to resources for those living in poverty. The mining sector increasingly recognises that it must work differently to how it often has in the past, and that there is potential for it to make a huge positive contribution to delivering the SDGs.

A changing industry requires a changing workforce – in turn, this requires a change in the mindset of students and academics alike. Raising students' awareness of development themes at the start of their careers may spark a flame that could persist throughout their working lives. Greta Thunberg exemplifies the impact a single student can have. Collectively, a generation of sustainability-driven geoscience students could make an enormous difference across a range of sectors. But how can this be achieved?

Geoscience programmes provide many opportunities to develop students' complementary and transferable skills, whether considering health and safety, access to water resources, environmental protection or the need for effective cross-cultural communication. Geology for Global Development (GfGD), a UK-registered charity, promotes the inclusion of such skills in university programmes and raises student awareness of the vital contribution their skills can make to sustainable development. Increasingly, interdisciplinary research on resource topics encompasses development themes such as gender, environmental protection, climate change and human health and wellbeing. Dissemination of insights from such research to students, and through them to the resources sector, can have a transformative effect.

Early-career and student voices have already begun to make a difference. Increasing attention is being paid to development topics in universities' natural resource modules, responding to the demands of industries being shaped in turn by a new generation of socially and environmentally aware geoscientists. Students can advance the development agenda further by asking provocative questions, and by stimulating evidence-informed debate among their peers and public audiences through social media.

They can also influence global sustainability discourses more directly – this may require them to show up in places they see as 'off-limits' or complex to navigate, but with support, their perspectives have the potential to be heard and acted on by policy-makers, UN agencies and wider civil society. The annual UN Forum on Science, Technology and Innovation for the SDGs provides one such opportunity. In 2019, GfGD led an international delegation of student and early-career geoscientists to this forum. Working with the UN Major Group for Children and Youth and student engineers, social scientists, medics and others, they helped to shape interventions regarding environmental education, the natural resource requirements for decarbonisation technologies to be deployed at scale, and the potential social and environmental impacts of this resource use. Their engagement resulted in the inclusion of key messages in the meeting outputs, later presented at the High-Level Political Forum on Sustainable Development. The wider significance of their engagement, however, is in demonstrating to the geoscience community that the best way to shape a conversation is to be part of it. Geoscience students and early-career professionals 'showed up', integrated their knowledge of geoscience into interdisciplinary statements, and in doing so caught the attention of diverse stakeholders in the room.

Father Nicolas Barla.
"You Are Dust, and Unto Dust You Shall Return."

On Ash Wednesday the Christian priest while putting the ash on foreheads reminds the faithful that "you are dust, and unto dust you shall return". It is a warning. While reflecting and writing for 'Of Earth, For Earth', I visualize the miseries and catastrophes inflicted on the world by the extractive industries and other development activities.

The universe is beautiful and rich, and includes the deserts that are beyond words! Every living being has the right to survive on earth. But we human beings, being the intellectuals and rational, we are trying to exploit other beings. The hill, earth, soil, forest, water, air, animal, they are also special beings. We human beings, because of our selfishness and greed, we extract and destroy the entire land: the forest, the water and other living beings. We human beings do not feel nor understand what happens to the environment while we extract. Our interest is only to earn revenue. Who thinks about sustainability?

The mining sector destroys nature and community. Those who are involved in extractive industries, most of them have no conscience about harm to the environment or local people. In the process of the extraction at the bauxite mines in Niyamgiri, Vedanta Resources was accused of several abuses toward the nearby Dongoria Kondhs, one of the ancient Indigenous communities in Niyamgiri . Thanks to environmentalists, and human rights activists, the Supreme Court of India supported the Gram Sabha (Village Assembly), who said 'No' to Vedanta, and the corporation had to leave the area in 2013. (Supreme Court judgment dated 18/04/2013 in W.P.(C) No.180/2011 Orissa Mining Corpn. Vs. MoEF & Ors.)

But in recent times some of us are experiencing catastrophe: the earth cracks; clouds blast and winds whirl; heavy rains and snows fall; waters rise from melting snow, in floods, tsunami, and the sea in the coastal belt; unbearable heat and fires; coronavirus or other pandemics. Why are all these things happening? Often, we do not want to question or reflect upon them! Often we are not affected, but others are affected - we do not care!

In spite of a high percentage of GDP (Gross Domestic Product), why in our cities are there so many beggars or sad peoples or migrant workers? Why are some are so rich and some are so poor? Why are so many peoples sick with new diseases? We do not want to ask these questions. No corporation is ready to find its solutions. Most of the corporations have no heart pangs, but go on extracting and destroying the earth and earning money.

I am an Indigenous Person from the 'Oraon' group close to Rourkela Steel city in Odisha, India. I was told by the late Mr. Gregory Ekka that "there were 93 Indigenous peoples' hamlets with over 6000 families in these areas". The government of India assured the Indigenous peoples that they would have good jobs, housing, roads, electricity, schools, hospitals … and they permitted the factory in 1954-1956. "We were not given jobs, because we had no engineering degree; no office jobs for we were uneducated, so we were totally cheated and betrayed. So it was also in the mining areas in Barsuan in the forest. There were many Indigenous Peoples' villages living in the forest. When the mining started they were displaced without anything." The tears used to roll from the eyes of Mr. Ekka while narrating all these miseries. He was a Landlord of over 100 acres of agricultural land. Today he is almost homeless, penniless, almost a beggar. Like him, thousands of families are homeless, many of them living in the slums. Many families disappeared. The hamlets are gone.

These types of occurrences are common all over the mining and industrial areas of the country and the world. Do these governments, corporations or business houses ever think of these unfortunate Indigenous / Displaced Peoples?

Wait, the time is near, when crony capitalism and the rosy life may turn upside down. As I see it, if you and I are not saving the earth soon, the days are coming when "you and I are dust, and unto dust we shall return."

Julian Allwood.
Mining with Zero Emissions.

We've got 29 and half years left to transform the world to operate with zero emissions - that's what scientists tell us will be safe, it's what social protestors are asking for, and increasingly it's written into our laws. Finland is committed to zero emissions by 2035, and following a change to the Climate Change Act proposed by Theresa May virtually in her last week in office, and passed unanimously through both Houses of Parliament in June 2019, in the UK we're committed to zero emissions by 2050.

With a little political caution, the Climate Change Act actually refers to "net zero" emissions, suggesting that there might be some meaningful "negatives" so that we can continue with some emissions. Unfortunately, this utopian dream isn't true - because all the apparent negatives either require an excess supply of biomass or an excess supply of renewably generated electricity but we have neither, and nor is there any prospect of having them in under 30 years. So "net zero" actually means "Absolute Zero" - and the research project I lead put out a major report with this title at the end of 2019 describing what this means.[1]

For miners and the mining industry there are two important consequences of our legally binding commitment to Absolute Zero. Firstly, the operations of mining - the yellow vehicles, crushing and grinding, transport and so on must all be electrified, with the electricity coming from non-emitting sources. That's largely feasible with existing technologies, although at present we don't have any electrified long-distance shipping. But secondly, the demands for mining products are going to change. Demand for the technology-metals associated with renewable power and electrification will increase significantly - just think of the impact of electrifying all the world's cars and domestic boilers, for a start. But meanwhile, demand for iron ore and coking coal will collapse, because we won't be able to operate blast furnaces that inevitably release greenhouse gases as part of their normal operation.

These two effects will create rapid and big change in the mining industry, but there's a third opportunity which is going to become very prominent in the next couple of decades. Up to now, miners have thought of themselves as people who dig new ores out of the ground, and then prepare them for processing. However, as we turn to an era powered only by electricity, there will be a huge growth in our interest in recycling, particularly of metals. For most metals, recycling can be powered by electricity and is much more energy efficient than making new metal from ore. The challenge is quality - which depends on the purity of the "feedstock" going into the recycling process. At the moment, that feedstock is relatively low quality - it contains a mix of lots of different metals - so the products of recycling are generally less good than the products made from new ores. But all that's required to sort that out is to be much more precise in separating out the different metals before they're melted.

And which industry is really good at separating out different components from a mixed stream of feedstock? The mining industry! So as rapidly as miners hang up their hard hats and clean up their grimy faces, as their old-style iron ore mines close, they could put on new white lab coats and get to work applying exactly the skills they have already developed, to purify streams of used metal. The colliery bands of the future will be playing with all the style and emotion we've known for centuries, but they'll be doing it with bright trumpets of the highest quality recycled brass!

[1] www.ukfires.org/absolute-zero

'Light Goes In' 60cm x 60cm, glass, solar cells, electronic components, 2019 Chloë Uden and The Art and Energy Collective.

Kathryn Sturman.
On the distance between researchers and mine-affected communities in Australia.

"Come in under the shadow of this red rock"
TS Eliot, The Wasteland

I came into Australia to study mine-affected communities, as a migrant into an economy of migrants, to a university full of migrants, self-styled as 'global citizens'. I wanted to understand the relationship between people and rocks. I still do. Now that citizens of the global village are experiencing our own forced resettlement under pandemic travel restrictions, it is a good time to reflect on how we do our research.

I've realised that the more we roam places to compare mining impacts the less we can know about those who live their lives in one place. What it is like to live under the shadow of a large-scale mine in a remote place? How would I know? We can visit these communities, interview individuals or run focus groups. We can 'hang out' like anthropologists or hassle people with questionnaires. We get the questions wrong and struggle with the answers.

The scale of mining in Australia is as overwhelming as the distances between urban, rural and remote communities. Between migrants, descendants of migrants and indigenous peoples. Between iron ore, coal, copper and uranium mines. How can a university degree be used as a stake to claim a part of this red earth and a right to ask questions of the locals? When I can get over that thought, there are some thoughts worth sharing.

Someone in East Arnhem Land shared that she has a 'love-hate relationship' with both her Yolngu community and the mining company she worked for. The company has been mining bauxite there since before she was born and will soon leave. As for the community: things can't go back to the way they were before the 1970s and they can't stay the same as they are now. But what's next?

There is a new indigenous-owned bauxite mine called Gulkula, a first in Australia. We get a tour and collect quotes from the manager: "At some point we realised that if our red earth has value to be dug up and shipped away, then it might as well be us who do the digging and the selling". Fair enough, as they say here. I think I get it, but cannot be certain.

Opposite page: '50°13'18.5"N 5°14'48.3"W Carn Brea, Cornwall.
Fine grained vein in coarse granite' Heidi Flaxman (detail).

Lucy Crane.
If You Care About the Environment, You Have to Care About Mining.

I care deeply about the environment, which is why I work in the mining industry. Do you think that makes me a hypocrite? A lot of people do, but I'm getting fed up with it. Mining and its products are vital for a civilised society – without it we can't build hospitals, roads, houses; we can't use the internet or fly around the world or post a photo on Instagram. The industry is fundamental to our daily lives, yet so many people have a negative view of it. Is this because there's a disconnect between what we consume, and where it has come from? There seems to be a lack of understanding about how crucial mining actually is: we need these raw materials, so we need to care about how and where they are being extracted.

If something hasn't been grown, then it's got to have been mined: everything we use in our daily lives has been produced from the ground in one way or another. I think we're starting to understand the impacts of what we're consuming when it comes to food: people understand that buying avocados from abroad might not be the best choice for the environment (not that it stops too many people from doing it!), but there is an understanding that local organic produce is probably preferable to something which is produced by intensive farming then shipped around the world before it lands on our plates. Arguably the mining industry has these same potential supply chain and environmental issues, yet amplified.

For example, the mobile smartphones in our pockets contain nearly two thirds of the elements of the periodic table within them. Just imagine how many mines these have all been sourced from, and the impacts associated with their extraction! As the middle class population of people grows around the world, in China and India and Africa, the demand for items such as cars and phones is rapidly increasing – and so is the need to extract more and more raw materials to build them from. Who are we to say that this mining isn't necessary, and that

people can't aspire to brick built houses, laptops and cars? However, it is not just luxury items that are increasing this global demand for raw materials. The need to produce low carbon technologies to combat climate change is having a significant impact too.

To put this increasing need for raw materials into perspective, in the past 5,000 years we have mined approximately 550 million tons of copper across the world. We need to mine that same amount in the next 25 years, just for use in low carbon technologies such as solar panels and electric vehicles. A typical 3MW wind turbine has nearly 5 tons of copper wiring in it, along with 2 tons of rare earths and 1,200 tons of concrete. And that's just one wind turbine! I think there's such a disconnect in society between what we're consuming and where it all comes from. If you care about the environment, then you have to care about mining too. It is time we started thinking about how we can extract these resources as efficiently and as responsibly as possible. This responsibility lies with all of us – as consumers, we have a significant amount of power. Every purchase we make is a vote with our wallets.

As geologists, we are equipped with a toolkit of skills to understand how the Earth works. This allows us to unravel the Earth's past and learn how previous climate events have impacted the world; it enables us to use the fossil record to understand evolution; and importantly it enables us to find energy resources and raw materials vital for modern society. Responsible mining is a key part of the solution to moving away from our reliance on fossil fuels and combatting climate change – and we need people who care about the environment to be leading the charge on this.

Gareth Thomas.
To Be A Miner...

Mining for me started as a boy following in my
Father's and Grandfather's footsteps, initially in
open cast and later in underground mines.

Mining has always fascinated me, the fact that
material extracted from the ground could be
used to create so many products.

This went on to determine my direction in life.
It taught me to strive for my ambitions and
make them a reality. The mining communities
that gave life to me were themselves born out of
mining and the need for coal, be it for electrical
power, for locomotives, for steel making, for
the manufacture of sugar and so much more.
This helped fuel the industrial revolution and
drove innovation beyond our wildest dreams.
From Trevithick to the Stevensons to Isambard
Kingdom Brunel to the Wright brothers, Steve
Jobs and Bill Gates, all owe their success to the
materials that were first mined - then processed,
moulded or burned - allowing them to create
the engineering marvels that have fuelled
human expansion.

So what does mining mean to me? It means
that we should not forget who we are, we
should not forget where the materials come
from that provide us with everyday things that
we take for granted. From a simple needle for
sewing your clothes to a piece of glass, to the
space shuttle, the materials that allowed these to
be created was first mined.

None of this would have been possible if the
miner did not exist.

To be a miner, you need to be determined
and stubborn but smart. You need to respect
your surroundings and do your best to protect
others as well as yourself. But above all, you
need to work as a team for nothing is achieved
on your own.

Luís Lopes, Vitor Correia & Stephen Henley.
Insect-sized Robots in the Invisible Mines of the Future.

Since the Bronze Age, humankind has been using tools to get mineral raw materials out of the ground. At first it was antler picks and stone axes. The ancient Romans developed wagons and trams to carry larger amounts of ore out of underground mines. In Mediaeval times miners set fires against the rock face and doused them with water so that the thermal shock would crack the rocks – and a little later used explosives like gunpowder for the same purpose. Each innovation has taken the technology further away from sustainability and towards an efficient but mechanised and inhuman system, with ever-greater negative impacts on the environment and on the health and safety of people working in the extractive industries. What we are trying to do in ROBOMINERS is to develop technology that leads back towards more sustainable ways of extracting raw materials from the ground.

In recent years, many people are questioning the need for mining in a push towards a greener and cleaner future. The facts are that mining essentially provides resources for a better, safer and more positive future, and the industry is more conscious of environmental and social constraints, fuelled by the need to get support of local communities for mining activities. This pushes us towards more responsible mining, from employing more efficient ore treatment processes to reducing wastes or extending the lifecycle of mines. But there is an underlying condition supporting all these more sustainable steps: the development of innovative technology.

The ROBOMINERS[1] European funded project, is using converging technologies in robotics, miniaturisation, and cost-efficient drilling to create a robot-miner prototype for difficult to access mineral deposits. The creation of a small (ultimately perhaps insect-size) miner robot can change the current mining paradigm that is defined by the human scale (it's the human size that defines the dimension of galleries in underground mines and open-pits for surface mining). Overcoming the need for opening big galleries to extract only a few centimetres of a mineralised vein, for example,

will reduce very significantly the amount of waste generated. Furthermore, such miner robots will be able to access deeper deposits, and the opening of insect-sized galleries will have minimum impact on the surface. Imagine a swarm of these robots working cooperatively, and we will see a game-changing advance for mining.

Underground mines, with insect-size galleries, will be virtually "invisible". This has the potential to reduce in an amazingly significant way the environmental impacts of mining and to enhance the social acceptance of the industry. "Invisible" mines will recover useful materials in an optimised, integrated flowsheet, future-proofing any resources that are not of immediate interest, rather than discarding them as wastes. The unfolding of "invisible" mines will change traditional economic feasibility assessments, and call for the development of intelligent business models, capable of integrating a range of different value streams. It will also shift the skills and competencies of the mining workforce towards more complex cognitive categories with increased requirements in digital literacy, alongside a holistic understanding of the value chains that are using mining outputs.

However, the push for sustainable mining cannot come only from projects such as ROBOMINERS, that are shaping a new future. It also needs a joint effort from governments and civil society, encompassing all stakeholders from producers to consumers. It envisages the participation of the local community in the development process. It entails more than one innovative technology or solution. In fact, only with joint efforts is it possible to deliver sustainable mining for a sustainable future. Such is the vision supported by ROBOMINERS.

[1] robominers.eu

Facing page: 'Elemental Eidos (cathodoluminescence) #3', 20 x 20 inches Lumen/Digital as C-Type Metallic Archive print, 2020 Josie Purcell.

REMAKING

Of Earth, For Earth brought together perspectives that question the meaning of a mine. It represents the dialogues that are required to design a future that enables humanity to live well using sustainably acquired natural resources. Modern approaches to extraction of raw materials thus have to embrace the reality that they affect people with very disparate experiences and views. This final contribution to Of Earth, For Earth describes the challenges of supplying resources for society, and describes the approach taken by the IMP@CT consortium to find technological solutions that remake mining for a more sustainable future.

In 2019, the United Nations Environmental Programme produced its 'Global Resources Outlook' report. The report emphasised that the use of natural resources has tripled since 1970 and continues to grow. The report concluded that natural resource use must be decoupled from both the environmental impacts of economic activity and human well-being, as an essential step towards a sustainable future. The Earth is not running out of natural resources, but the stocks of 'good' metallic ore deposits are diminishing. Good stocks are the large and accessible ore deposits that contain a high concentration (or grade) of the commodities of interest. The largest, highest-grade and shallowest deposits have been depleted. Now, we must look to the rest.

Economic and regulatory frameworks support production of metals from large mines, where innovation enables mines to run at increasing efficiency using the economies of scale. This keeps production costs low, to underpin the rest of the manufacturing value chain. An adequate return on investment requires that the ore deposit support an extended life-of-mine. Decreasing cost per unit output with increased mine size favours the development of giant or 'world-class' deposits. But

large-scale mining of low-grade ores is energy-intensive and generates vast wastes.

An addiction to scale in the mining industry means that metals consumed in small amounts annually can be supplied by very few mines, with distribution through global transport networks. Should conflict or trade disputes arise in the locations of these mines, then the real or perceived threat to metal supply impacts the commodities markets. Price volatility in turn impacts the mining sector because the metal price dictates the cut-off grade (the concentration of commodity in the rock) below which it is no longer economic to mine. A low metal price reduces the economic feasibility of operations but a high metal price can open up short-term opportunities for all mined commodities. The commodities considered at critical risk of short supply include those used for the infrastructure of the green energy transition, the low-carbon transport fleet and modern digital communications.

There are existing and feasible technologies for selective mining of small ore-deposits by small-scale operations. There is a difference between mining of small deposits and small-scale mining operations. Small deposits are necessarily described using geological and metallurgical terms, in order to engineer mining and processing solutions. The practice of mining small deposits transects historical and contemporary narratives: Many of the opportunities for future mining lie in the historic mining districts that were never exhausted of mineral deposits, but that suffered the impacts of mining that was not environmentally or socially regulated. Scale describes the size and complexity of mining operations and their social and environmental impacts, both positive and negative. Small-scale mining will have smaller and shorter duration economic and environmental impacts than

M I N I N G

large-scale mining of low-grade ores. Opportunities exist to consider mining of small deposits as part of locally-diversified economies, remediated environments, and geographically-dispersed and secure raw materials supply. Small-scale mining is not supported by current trade, reporting and finance models. But increasing global demands for best practice, equitable distribution of opportunities and reduction of carbon emissions are external forces that may create a climate amenable to the expansion of mining of small deposits.

IMP@CT is an acronym for *'Integrated mobile modularised plant and containerised tools for selective, low-impact mining of small, high-grade deposits'*. The IMP@CT project partners investigated the potential to expand European capacity to access its regional resources, following from four key traits of the whole system:

1. Risk of raw material supply disruption can be reduced by accelerating the response of mining to market demand, facilitated by access to multiple small deposits in Europe.

2. A new mining paradigm is needed that does not rely on extensive investment and the economies of scale.

3. Energy demand and mine waste should be reduced by limiting the volume of rock removed from the ground and crushed.

4. Mining solutions should have a minimal footprint to support multiple land uses for high population density, and an ethical relationship with community. The project partners developed technological readiness for non-artisanal, sustainable small-scale mining that would be suitable for extraction of metals from small ore deposits in Europe, and interrogated the societal solutions that would

facilitate the mining approaches (see Olga Sidorenko, this volume).

The partners established boundaries within which to discuss what constitutes a 'small deposit', prioritizing the notion of 'high-grade'. The researchers aimed to understand how small a deposit can be and remain viable to mine, and to gain some appreciation of the extent of opportunity for mining of small deposits in Europe. They adapted an existing search engine[1] that enables any end-users to identify and locate small, high-grade complex deposits. Europe is a region with a varied geological history, such that there are many and diverse small, high-grade ore deposits.

A new selective underground mining tool was designed and developed with small size to reduce the size of underground workings and leave rock in place. The cutting head was designed to control the size of particles at the rock face, reducing the need for subsequent crushing of rock. Waste material was further separated from metal-rich rock, prior to crushing by ore sorting technology. A portable processing plant was designed and developed to adapt to different ore deposits. All of the technological infrastructure was designed to be mobilised in 20-foot long containerised modules, so that it could be transported to ore deposits located in rugged terrains and rapidly deployed in a configuration that has a small environmental footprint. The component parts were integrated and tested on two mine sites in the west Balkans in 2019. The first deployment was at the Olovo lead mine in Bosnia and Herzegovina, where all processing occurred on site. The second deployment was at the Velicki Majdan processing site in Serbia, to process antimony ore from the nearby Zajača mine.

Energy consumption reduction by selective mining, crushing and processing meant that the IMP@CT system could be powered as consistently and more cost-effectively in the Balkans region by renewable or combined renewable-diesel energy provision, than by fossil fuels alone. This is a significant development since the mining industry accounts for approximately 10% of world energy consumption. The environmental loads of mining are not limited to carbon emissions. Multiple partners investigated the processing methods, the treatment of industrial water and the opportunity to use IMP@CT solutions to process and thereby remediate legacy wastes.

In summary, the IMP@CT project showed that technical solutions are available now to accelerate decoupling of mining from negative environmental impacts and human well-being, in order to realise ambitious 2050 targets (see Julian Allwood, this volume). The solutions will bridge the gap until more future-facing innovations are developed in the coming decades (see Luis Lopes et al, this volume).

However, the IMP@CT project also showed that small-scale mining solutions cannot succeed without reshaping of existing customary patterns of the governance of socio-economic impacts of mining. There are four key issues that must be addressed for the future development of small-scale mining on small, high-grade ore deposits in Europe:

1. A positive relationship between mine and community must be prioritized, even in short-duration operations, underpinned by local micropolitical and cultural understanding.

2. Permitting and licensing processes must continue to protect environment and society, while enabling mine operators to react swiftly to market opportunities.

3. The role of short-duration mining must be placed within a diversified local or regional economic base, to protect the community from the withdrawal of the mining industry.

4. Further innovation in handling of waste is required, since short duration mining operations must not create a long-term environmental legacy.

As a final note, the Covid-19 pandemic has hit the mining industry hard because of decreased demand leading to plummeting commodity prices, on a scale equal to that of the 2007-2008 global financial crisis, and the necessary closure of mines due to national and regional lockdowns. Small deposit mining by small-scale operations may assist the mining industry to regenerate after the pandemic since it requires lower investment. This will be important in a market of low prices, low production and potential over-supply. Even in the fiercely competitive economic climate that the mining industry periodically operates, it is possible to build back better.

Kathryn Moore, and the IMP@CT Consortium.

[1] EU-Minerals Knowledge Data Platform

The University of Exeter, Camborne School of Mines, United Kingdom
emps.exeter.ac.uk/csm

BRGM (Bureau de Recherches Géologiques et Minières) France
www.brgm.eu

Mineco Ltd, United Kingdom
www.minecogroup.com

RWTH Aachen (Rheinisch-Westfaelische Technische Hochschule Aachen), Germany
www.rwth-aachen.de

Imperial College London, Royal School of Mines, United Kingdom
www.imperial.ac.uk

Extracthive, France
www.extracthive-industry.com

EPSE (Global Ecoprocess Services OY), Finland
www.epse.fi

Cymru Coal Ltd (Metal Innovations), United Kingdom
www.metalinnovations.co.uk

Rados International Services Ltd, United Kingdom
www.radosxrf.com

University of Eastern Finland (Ita-Suomen Yliopisto), Finland
www.uef.fi

Photo: Lars Barnewold.

CONTRIBUTORS

SECTION ONE

KATHRYN MOORE

Kathryn Moore gained degree qualifications in geology (BSc), experimental petrology (PhD) and archaeology (Dip) from the University of Edinburgh, University of Bristol and National University of Ireland, Galway. She led the Magmatic Studies Group at the National University of Ireland, Galway from 1999 and moved to the Camborne School of Mines in 2012, as lecturer in Critical and Green Technology Metals. She participates in and supervises research relating to alkaline rocks, carbonatites, ore deposits (particularly critical metals), and small-scale mining. Kate is the Project Lead for IMP@CT (Integrated Modular Plant and Containerised Tools for selective, low-impact mining of small, high-grade deposits). k.moore@exeter.ac.uk www.impactmine.eu

BRIDGET STORRIE

Bridget Storrie is a PhD candidate at the Institute of Global Prosperity at UCL, supervised by Professor Dame Henrietta Moore. Her research focus is the relationship between natural resources, conflict and peace. She has an MA in Russian (St Andrews University), a Masters in Peacebuilding and Reconciliation (Distinction) from Winchester University and is a trained mediator (Justice Institute of British Columbia). Bridget worked as a foreign news producer for ITN in Moscow in the early 1990s. She is married to a mining engineer and has lived and worked in Namibia, Alaska, Australia, Canada, Mongolia and Serbia. bridgetstorrie@icloud.com

DANA FINCH

Dana Finch has worked as a project manager on several European funded research projects since 2006, at Imperial College London and King's College London before moving to the University of Exeter to become the project manager of the IMP@CT Project. She is also a practising artist, a graduate from Dartington College of Arts, and exhibits regularly in the UK. She initiated and co-curated the Of Earth - For Earth exhibition and is a co-founder of Deep Earth Synergies, an arts hub dedicated to bringing artists and mining professionals together to examine novel solutions to age-old problems. www.impactmine.eu. danafinchartworks@gmail.com

DAN PYNE

Dan Pyne was born in London, and studied Scientific Illustration at Middlesex University. Since 2010 he has lived and worked in Cornwall teaching on workshops and projects across the county including at Tate St Ives. He currently works at the Newlyn School of Art and is a recent Vice-Chair of the Newlyn Society of Artists.
His practice is principally an exploration of materials and experiments in process that teeter between structure, control and self creating organic objects. The works often inhabit a space between sculpture, installation and painting, shaped as much for their haptic qualities as the visual. danny.pyne@btinternet.com

CARLOS PETTER

Carlos Petter has a Bachelor's degree in Mining Engineering from the Federal University of Rio Grande do Sul, Brazil (1986), and has a PhD in Techniques et Économie de L'Entreprise Minière by École des Mines de Paris (1994). He was a post-doc at École des Mines d'Alès (2002), in mineral fillers on polymers, and the University of Exeter (2018) in Mining and Sustainability. Currently he is Professor at UFRGS.He has experience in mining engineering, with an emphasis on Economic Evaluation, and experience in projects with companies in the mining and industrial minerals sector (Vale, Imerys, PPG, Renner-Herrmann, Braskem). He has 50 publications in journals and 50 publications in congress proceedings. cpetter@ufrgs.br

ALAN SMITH

Alan Smith is an artist working in video, sound and live events. Since 1994 has been co-founder and creative director of Allenheads Contemporary Arts. His practice is fuelled by the landscape of the North Pennines, exploring disused lead mines with a mine explorer and a biologist to submerge canvases in disused reservoirs and air shafts which collected sediments and fungi, the source of his visuals. Later, the 'experiential' became his primary concern. 'Parameter' seven artists spent 24 hours in a subterranean cavern explored non-verbal experience. 2017 this developed into 'Chthonic' when three artists and one engineer lived for 72 hours underground. His practice transfers across urban and rural environments and has led to research projects at his base at ACA https://www.acart.org.uk/art-projects-aca and overseas at Nida Arts Colony, Lithuania and Massey University, New Zealand, where he explored links between astronomy and the arts, this work continues at ACA. www.acart.org.uk/alan-smith

LOUISE K. WILSON

Louise K Wilson is a visual artist who makes installations, drawings, live works, sound and video works. She frequently involves the participation of individuals from industry, museums, medicine and the scientific community in the making of work and previous associations have included Newcastle International Airport (air traffic control), Montreal Neurological Institute, the Science Museum, London and the Yuri Gagarin Cosmonaut Training facility, Moscow. Recent exhibitions include Meetings Festival (ET4U, Denmark, 2019); Thackray Uncovered (Thackray Medical Museum, 2017); Submerged: Silent Service (Ohrenoch, Berlin, 2015); and Topophobia (Danielle Arnaud Gallery, London; Bluecoat Gallery, Liverpool and Spacex Gallery, Exeter, 2012). Her programme Cold Art - exploring artists' fascination with Cold War sites - was broadcast on BBC Radio 4 in 2018. She is a lecturer in Art and Design at the University of Leeds and lives in West Yorkshire. lkwilson.org

DYLAN MCFARLANE

Dylan McFarlane was a mining engineer with a professional career developing and operating small-scale gold mines spanning 14 years. His technical expertise included geological exploration, mineral processing, mine reclamation, and tailings/waste management. He worked with Anglo

American on their proposed Pebble gold-copper mine project in Alaska and at Los Bronces copper mine in Chile. He gained his MSc in Mining Engineering from the Camborne School of Mines (CSM), and an MA in Sustainable Development from the University of St. Andrews. Dylan was project manager for the EU-funded HiTech AlkCarb research project and involved in teaching and research. Most recently he worked for PACT, a non profit international development agency.

ADELE ROULEAU
Adele has a background in engineering and law. She has a fascination for the synergies between the mining industry and society, as well as its evolution throughout history. From the transition to a low carbon economy, to the opportunities for mining in space, the mining industry will undeniably influence our world as well as be influenced in the centuries to come as it has done in the past.
Adele is currently specialising in international mining taxation and law.
ar15776@my.bristol.ac.uk

JOSIE PURCELL
Josie Purcell is based in Cornwall. Her photographic practice predominantly looks at the human impact on the natural world through the use of alternative and camera-less photographic processes. She set up her participatory photography project, ShutterPod, in 2014, and gained a distinction in her MA in Photography in 2019. Josie has exhibited internationally with organisations such as Shutter Hub and with Rome Art Week.
jepurcell@btinternet.com

JACK HIRONS
Jack Hirons lives and works in London. Jack studied BA (hons) Photographic Arts at the University of Westminster. Previous exhibitions include Bone Black, Backroom Gallery, London (2019); All or Nothing, Lungley Gallery, London (2019); Material Light, Kochi Biennale, India (2017); Material Light, Kulturni Centar Beograda, Serbia (2015)
hironsjm@gmail.com

DOMINIC ROBERTS
Dominic Roberts has been working in the Balkans on and off for the whole of his career and exclusively for the last 13 years. As Head of Corporate Affairs of Adriatic Metals PLC he has responsibility for the permitting and licensing of the Tier 1 Vares poly-metallic project in Bosnia and the recently acquired Raska poly-metallic project in Serbia. Prior to joining Adriatic Dominic was the Chief Operating Officer of Mineco, the largest base metal mining group in the Balkans and he opened the Olovo lead mine in 2019, the first new Bosnian mine to be permitted in a generation. He was also additionally responsible for the optimisation and integration of Mineco's five mines in Bosnia and Serbia. Prior to entering the mining industry Dominic served with the British Army, completing operational tours, amongst other places, in both Bosnia and Kosovo. Believing in both the potential and the absolute future requirement for European metal mining Dominic has been a driving

member of the IMP@CT consortium from fruition.
dominic.roberts@adriaticmetals.com

OLGA SIDORENKO
Olga Sidorenko holds her Master's degree in Environmental Policy and Law from University of Eastern Finland. Her areas of focus include social scientific research on extractive industries, company-community relations and sustainability challenges of mining development. In the IMP@CT project, Olga has worked on local community perspectives and factors of social acceptance of small-scale mining. Her previous work in the University of Eastern Finland examines the challenges of sustainable mining governance in post-socialist countries.
olga.sidorenko@uef.fi

PENDA DIALLO
Before joining CSM, Penda was Senior Resilience Advisor at CARE International-UK, providing technical support and project development assistance on resilience building projects in various CARE country programmes including Niger, Nepal, Somaliland, South Sudan, and Haiti. Before joining CARE she was a livelihoods and governance advisor, where she worked on natural resource management projects in Burkina Faso, Sierra Leone, Guinea and Cameroon. Previously she had worked as an environmental consultant for AECOM, and in research and communications roles for the Conflict, Security and Development Group in King's College London, Helen Keller Worldwide in Guinea. She has recently published a book on her research into bauxite mining, Regime Stability, Social Insecurity and Bauxite Mining in Guinea.
p.n.diallo@exeter.ac.uk

FRANCES WALL
Frances Wall is Professor of Applied Mineralogy is an expert in critical metals and responsible mining research. Her background is in carbonatites and their ore deposits including, especially now, critical metals, working first at the Natural History Museum, London and then at Camborne School of Mines where she held the role of Director for a time. She was part of the Natural Environment Research Council Expert group for the Mineral Resources Programme on Security of Supply of Minerals (2011-13) and currently leads a number of large research projects with Universities, geological surveys and exploration companies. In 2019 she was awarded the Geological Society of London's William Smith Medal, the first ever female recipient.
f.wall@exeter.ac.uk

SECTION TWO

HENRIETTA SIMSON
Henrietta Simson is an artist whose work explores the landscape image through its historical and cultural development, and its current definition within a digital context framed by ecological crisis. Drawing from late medieval and early Renaissance imagery, she presents an idea of landscape that challenges its designation within human/nonhuman dichotomies and that facilitates a critical questioning of the

visual structuring of space. She completed an MA at the Slade School of Fine Art in 2007, and a practice-related PhD in 2017: Landscape After Landscape, Pre-Genre Backgrounds in a Post-Genre Digital Age.
henriettasimson@yahoo.co.uk

DOMINIKA GLOGOWSKI
Dominika Glogowski has a broad experience and professional background in arts management, art history, and visual arts. She is the founder of the think tank artEC/Oindustry on mining, nature and the arts. She bridges theoretical and applied practices and focuses on the arts' inclusion in the industrial sector, STEM and life sciences. Tackling complexities and solutions of our modern way of living, Dominika initiates and designs interdisciplinary projects around the globe, as recently the art-science ClimArtLab with the Konrad Lorenz Institute for Evolution and Cognition Research in Austria and the art hub Deep Earth Synergies in Cornwall, UK, which she initiated and co-founded to explore cross-sectorial approaches for a participatory socio-ecological future in extractive areas.
www.artecoindustry.com office@artecoindustry.com @artecoindustry

JAMES HANKEY
James Hankey is an artist based in the South West, and has exhibited widely across the UK. His practice develops through photographic, performative and often absurdist processes of production that reflect on and conflate local histories and wider ecologies.
jameshankey84@gmail.com

KIERAN RYAN
Kieran Ryan (BSc, MSc, PhD, PGCE) has a research background in sedimentary geochemistry and engineering geology. He subsequently worked for 7 years as an environmental consultant focussed on land remediation, hazard assessment, site investigation for environmental impact assessment and renewable energy installations. Subsequently he became an educator and is now Head of Geography and Geology at Helston Community College.
kieran.mryan@gmail.com

ALISON COOKE
Alison Cooke is a London based ceramicist. She works with clay dug from interventions such as mining, engineering or scientific research. Her work revolves around the layers of history below a location, with a particular focus on hidden or underground networks where histories might relate to, and impact on our lives today and in the future. The materials are used in their natural state and fired at an extreme range of temperatures. Alison's work is self-initiated, project based and often includes public engagement and involvement of other artists. She is co-founder of ceramics collective the Associated Clay Workers Union (ACWU).
alisoncooke.co.uk

KARIN EASTON
Karin Easton served as Honorary Secretary for Perranzabuloe Museum in Perranporth and still serves as a volunteer on the Museum Management Committee. Since retirement as Deputy Head of a Cornish primary school, she obtained an MA in Cornish Studies with the Institute of Cornish Studies, University of Exeter. She is currently President of the Federation of Old Cornwall Societies of which there are over 40 across Cornwall.
perranzabuloemuseum.co.uk

CHRIS EASTON
Christopher Easton was a founder member and is a Trustee of Perranzabuloe Museum in Perranporth and has been Chairman of the Management Committee. He is President of Perranzabuloe Old Cornwall Society. As a keen beachcomber he is proud of his collection of sea beans and transatlantic flotsam. The latest edition of A Flora of Cornwall mentions some of his rarer finds of disseminules or sea beans. He produces artwork from the plastic he finds on Perranporth beach. Some of his public art highlights the message of plastic pollution.
perranzabuloemuseum.co.uk

NIC BARCZA
Nicholas Adrian Barcza studied sciences including geology, and participated in sports, art and drama at St Stithians College. He completed BSc, MSc and PhD degrees in metallurgical engineering at the University of the Witwatersrand in South Africa. He developed an interest in the mineral chromite and studied its reduction with carbon to produce chromium alloys used together with nickel in stainless steel that we use in our everyday lives in so many ways. In the late 1970s he worked in the Cotswolds with applied research into the use of thermal plasma arc technology that has been developed at Mintek. He has an ongoing global interest in and involvement with the development of natural resources and recycling of materials and sources of energy to minimise environmental impact and support the sustainable production of materials that contribute to our quality of life in a responsible manner.
n.barcza@btinternet.com

NIC CLIFT
Nic's broad and holistic interests are not limited by his being a Veteran of the Mining and Metals Extraction/ Refining/Recycling industries, with extensive international experience in: Operations; R&D; Consultancy; Management; including as CEO and MD of two ASX listed companies; Community and Government Relations. A Senior Industry Fellow of RMIT University, he has a special interest in the implications of the adoption of disruptive technologies, and the >15 year horizon of politics "beyond the election cycle". "Living the values of working harder and smarter to create equal opportunity and achieve sustainable development" Nic was Director General of CBG from 2004-7
nicclift@icloud.com

DJIBO SEYDOU
Djibo SEYDOU BSc Mining Engineering, MSc Environmental Engineering and Safety in Mining; Head of Mining Nuclear and Radiological Safety and Security Division in the Mining Environmental Department, Ministry of Mines, Niger. He has a long experience in mining environmental

management having held several positions within the Mines Directorate: Head of Classified Establishments Service, Head of Radiation Protection Service and Head of Mining Environment and Classified Establishments Division but also for having carried out several monitoring and control missions on the uranium mining sites of Arlit, Akokan, Azelik; coal mining site of Tchirozerine and gold mining site of Samira.
seydou_djibo@yahoo.com

NAOMI BINTA STANSLY
Naomi Binta Stansly is a human rights defender in Niger, and is president of the NGO Tedhelte Entre Aide Niger and Vice Coordinator of the Network of Organizations for Transparency and Budget Analysis (ROTAB) Niger. Tedhelte works on the economic, social and political rights of women and young people. ROTAB is a coalition of civil society organisations in Niger, which promotes transparency throughout the activities related to the extractive industries. She is mother of four children.
rotabniger.net naomis03@yahoo.fr

RICHARD MARTIN
Dr Richard Martin is an anthropologist at The University of Queensland. He has undertaken anthropological research in Australia and the United States. His research focuses on Indigenous and settler histories in post-colonial settings. He has published a range of scholarly articles about Indigenous people and culture, and is the author of The Gulf Country: The story of people and place in outback Queensland (Allen & Unwin, 2019). He has also authored numerous expert reports about Indigenous land negotiations, and given evidence in the Federal Court of Australia on behalf of Aboriginal people claiming rights in land and waters.
r.martin3@uq.edu.au

OLIVER RAYMOND-BARKER
Oliver Raymond-Barker works with the mechanics and alchemy of photography to make images, objects and structures that expand upon what photography is and can be.
Working predominantly with alternative analogue techniques he uses photography as a tool to uncover imagined narratives, unseen processes and underlying systems.
Recent exhibitions and displays include PhotoIreland (2019), IED Madrid (2019), Belfast Photo Festival (2019), UNSEEN photo festival, Amsterdam (2018), Pic.London (2018), Newlyn Art Gallery, Cornwall (2017) & Four Corners Gallery for the London Pinhole Festival (2017). From 2014 - 2015 he was creative Associate at the University of Exeter's Environment and Sustainability Institute (ESI) where he collaborated with Dr. Caitlin De Silvey and Dr. Chris Bryan on The Natural Alchemy project.
He was recently selected to participate in the Peer Forum programme at The Photographers Gallery, London and has delivered talks and workshops for a range of institutions such as UNSEEN (Amsterdam), Cove Park (Scotland), YATOO (South Korea) & London College of Communication (University of the Arts London).
oliverraymondbarker.co.uk

CAITLIN DESILVEY
Caitlin DeSilvey is a Professor of Culture Geography at the University of Exeter and co-manager of the Creative Centre at the Environment and Sustainability Institute (ESI), University of Exeter. She is coauthor of Visible Mending and coeditor of Anticipatory History.
c.o.desilvey@exeter.ac.uk

SECTION THREE

GILL JULEFF
Metals and minerals are embedded in my DNA. Brought up on a farm in Cornwall in a family with as many miners as farmers, I was always more interested in the names of ruined engine houses dotted across the landscape than learning how to keep chickens. I was captivated by stories of mines in Brazil, South Africa, Australia and California where relatives went to seek their fortunes. I first left home for Art College where, primarily through social contacts, I discovered archaeology. Good fortune gained me an undergraduate place at the Institute of Archaeology in London where the study of archaeometallurgy captured my curiosity. An opportunity for a year's work experience with UNESCO in Sri Lanka turned into 13 glorious years of adventure, including PhD research and the discovery of a novel 1st millennium iron smelting technology that utilised powerful monsoon winds to create high-carbon steel. Returning to UK and Exeter University, I continue to work on iron in Asia with research colleagues in India, China and Japan but I have also rekindled my local interests in both the archaeology of mining and the social history of mining communities.
Cassia Johnson
g.juleff@exeter.ac.uk

HEIDI FLAXMAN
Heidi Flaxman is a mixed media artist using traditional craft processes and digital fabrication methods. Her practice investigates intangible landscape; through craft she transforms vast areas into tangible artefacts which encourage alternative visual exploration of an area.
"Through my creative practice I present an alternative view on a landscape through the coalescence of physical geographies and microscopic geologies which are largely unseen. Exposure of these elements is used to enhance a narrative of the landscape, its features and histories. Post-industrial landscapes, where topography has been altered as a result of the exploitation are inspiration for my work".
heidi@heidiflaxman.co.uk

ANSHUL PANERI
Anshul Paneri joined the University of Exeter's Renewable Energy department as a Postdoctoral Research Associate in 2018. His research focuses on the feasibility study of renewable energy systems (RES) for the IMP@CT project, where he has simulated several models of hybrid RES for small complex deposits in Europe. The overall finding shows that it is technically and financially feasible to use existing RES for the IMP@CT project energy demand with significant reductions in GHG emissions compared with conventional diesel generators.
a.paneri@exeter.ac.uk

CONTRIBUTORS

CASSIA JOHNSON

Cassia is an exploration and Quaternary geologist living on the West Coast of Canada. She has spent her career exploring for precious metals and gems, and surficial mapping in multiple locations around the world, from the arctic tundra of Nunavut to the tropical mountains of the Dominican Republic. After almost a decade of working as a professional geologist, she is following her passion and pursuing a PhD focused on artisanal and small-scale mining in west Africa at the Camborne School of Mines. A geologist gone social, she hopes to bridge the technical and social aspects of artisanal and small-scale mining. When she is not working on her PhD, you can find her on her yoga mat, lost in her journals and novels, or hiking and running in the wilds of British Columbia.
cj422@exeter.ac.uk

HEATHER WILSON

Heather Wilson is an artist printmaker. She graduated in fine art in 1984, and has a masters in illustration. She lives in Edinburgh, Scotland. She has exhibited widely in Scotland, the UK and abroad. Having spent 2 years, 1988-1990, printing Thomas Bewick (1753-1828) wood engraving blocks for an archive at Newcastle Central Library, in England, she started to make her own wood engravings and has continued ever since. She now likes to experiment with combining wood engraving, wood cut, mono-print and collage. Her work is based on personal experience, memory, myth and a sensitivity to land, sea, flora and fauna.
hettywilson@rocketmail.com

ALLIE MITCHELL

Allie Mitchell is an Environmental and Social specialist working with mid - junior tier mining companies around the world. Her time is predominantly spent engaging with communities close to mine sites, establishing a relationship with them to understand their perceptions and concerns of a Project, and to develop sustainable and mutually beneficial resolutions. After completing a degree in Geology and a masters in Management of Natural Resources, Allie noticed the gap in the mining industry where technical information is often not portrayed in an understandable manner to communities, this is therefore where a lot of her time and energy is focused.
allie@gfgd.org

JOEL GILL

Dr Joel C. Gill is Founder and Executive Director of the charity Geology for Global Development. For the last decade, Joel has worked at the interface of Earth science and international development, focusing on themes including sustainability, disaster risk reduction, equitable partnerships, and education for sustainable development. Joel plays a leading role internationally in championing how geoscientists can help deliver the UN Sustainable Development Goals, and is the lead Editor of an upcoming book on this theme. He was elected to the Geological Society of London Council in 2019 and to the position of Secretary, Foreign and External Affairs in 2020.
@JoelCGill

NIC BILHAM

Nic Bilham is a researcher at the University of Exeter, where he is working on responsible sourcing of minerals, the relationship between mining and the circular economy, and the challenge of assuring environmental and social impact standards across complex value chains and production-consumption networks. Until 2018, he worked at the Geological Society of London for over 20 years, most recently as Director of Policy and Communications. He has a longstanding interest in interdisciplinary approaches to global societal challenges relating to meeting our resource needs sustainably. Nic is chair of trustees of Geology for Global Development (GfGD), and an Executive Council member of the International Association for Promoting Geoethics (IAPG). He holds degrees in History and Philosophy of Science (BA, University of Cambridge) and Science and Technology Policy (MSc, University of Sussex). nb533@exeter.ac.uk

FATHER NICHOLAS BARLA

Nicholas Barla is from an Indigenous Tribe called 'Oraon' from Odisha India. After graduation in Commerce in Sambalpur University in 1984, he spent time in villages and realised the systemic exploitations that inspired him to study law. After a few years working in the courts, he experienced the complications and corruption involved in the justice delivery system that inspired further study for a Master's degree in Business Administration(MBA). He started empowering the Indigenous Peoples in the villages on their land, forest, socio, economic and cultural rights, and worked in the State of Odisha closely with the government and NGOs. He was nominated as an Advisory member in the Ministry of Rural Development (MoRD), Government of India where he was able to dialogue and mediate people's issues, both in the State of Odisha and at the National level.
In 2017 he was nominated as Steering Committee Member in UNESCO for Organizing the International Year of Indigenous Languages 2019 for Asia region. 'Basically I like to work together with poor, marginalized peoples for a just society and I am committed to the cause and it gives me joy'.
nickybarla@gmail.com

JULIAN ALLWOOD

Julian Allwood is lead author of the recent UK Fires report Absolute Zero, Professor Allwood's work focuses on material efficiency as a climate mitigation strategy, as well as whole-system analysis of natural resource systems. Professor of Engineering and the Environment at Cambridge University and author of the 2012 book "Sustainable Materials: with both eyes open", his research group investigates material demand reduction, novel material saving manufacturing processes and whole-system analysis of energy, land, water and material resource systems. Prof. Allwood is also Director of the UK Fires group on industrial resource efficiency, whose recent Absolute Zero report has been making waves in the policy sphere.
Professor Julian Allwood FREng, Professor of Engineering and the Environment, Cambridge University & Director, UK Fires
jma42@cam.ac.uk

CHLOË UDEN

Art and Energy, founded by Chloë Uden, is a collective of artists, technologists and makers based in the UK working exclusively at the intersection between art and energy. We have challenged the existing methods and processes that go into making silicon solar panels to push the aesthetics of the panels enabling us to use the materials that go into the construction of solar panels as materials for making art. We are very interested in the materials that go into the manufacture of solar cells come and we know that many of these are mined, and that we will need to continue to mine these materials in order to build a sustainable future. Our artworks embody the stories of the elements needed to build a healthier relationship with the earth.
www.artandenergy.org

KATHRYN STURMAN

Dr Kathryn Sturman is a Senior Research Fellow and Postgraduate Coordinator at the University of Queensland's Sustainable Minerals Institute (SMI). She has a PhD in International Relations from Macquarie University, Sydney and an MA cum laude in Political Studies from the University of Cape Town. Kathryn's research focus is on the political dynamics shaping the extractive industries globally, particularly in resource-rich developing countries of Africa and South East Asia. Before joining SMI in 2012, Kathryn was program head of the Governance of Africa's Resources Program at the South African Institute of International Affairs. She has conducted research and policy development in the minerals, oil and gas, and logging sectors. She has also worked as a senior researcher for the Institute for Security Studies and a speechwriter in the Parliament of South Africa.
k.sturman@uq.edu.au

LUCY CRANE

Lucy Crane is currently a Senior Geologist at Cornish Lithium and heavily involved the company's business development and expansion. She holds a degree in Earth Sciences from Oxford University, and a Masters in Mining Geology from the Camborne School of Mines. Lucy is a strong advocate for the standardisation of sustainable and responsible practices in mining, promoting these actions to the wider public, and is passionate about making the industry a more diverse place to work. Her actions in these areas are already being recognised and was awarded the Rising Star Award at Mines & Money (London) in 2019. Her 2019 TEDx talk on 'Mining our way to a low carbon future' has been well received.
l.crane@cornishlithium.com

GARETH THOMAS

Gareth Thomas is the Managing Director of Metal Innovations UK, with over 30 years experience in equipment design, manufacture, open cut mining and underground mining, starting in 1981. 1991 he founded SL Engineering Ltd, and designed the first hydraulic walls in the world for Brent Walker at the Cardiff International Arena, now known as the Motor Point Arena. These were displayed in operation to HM the Queen when she opened the arena in 1993. He has designed mines, river dredges, excavators and many more, and has travelled overseas, exporting skills and equipment. "It has been my life's ambition to bring work home to Wales and the UK, to at least play a part in providing work and income for the communities I grew up in".
gareth@metalinnovations.co.uk

VITOR CORREIA

Vitor is Secretary-General of the International Raw Materials Observatory and Past-President of the European Federation of Geologists. He founded and managed several companies working in geosciences, and he has over 25 years of experience in strategic management, innovation and organizational effectiveness. He began his career as a mining geologist, and he worked in mineral exploration, geological engineering and environmental geology in Europe, Africa and South America. Vitor holds a BSc in Geology and an MBA, both from the University of Lisbon. He is registered as a EuroGeol.
Robominers.eu

LUIS LOPES

Luís Lopes is a Portuguese geologist with an educational background in Earth Sciences and interests in sustainability, innovation and project management. He has a Master's degree in Geomaterials and Geological Resources, from the Universidade de Aveiro, Portugal. He is working with La Palma Research Centre in charge of communication and dissemination efforts as well as research roadmapping activities in many EU-funded raw materials related projects including the Horizon 2020 ROBOMINERS, UNEXUP and INTERMIN.
Robominers.eu luislopes@lapalmacentre.eu

STEPHEN HENLEY

After a conventional geological mapping and geochemistry PhD project on a highly mineralized area in Cornwall, UK, he has spent most of his career concentrating on computer applications in geology. Following several years with Australian and UK geological survey organizations, he founded the 'Datamine' mining software company in 1981, recognized by the Aberconway Medal of the Geological Society of London in 1992 and Queen's Awards for both export and technology. Since 1993 he has worked worldwide as an independent exploration/mining consultant, with extensive experience in Russia, as well as a research contract at CSIRO in Perth (Australia), on predictive exploration. In recent years he has worked mainly on EU projects including development of robotic technology for exploration and mining.
Robominers.eu

The IMP@CT project is funded by the European Union's Horizon 2020 research and innovation programme under grant No 730411 (2016-2020).
Of Earth, For Earth is an output of the IMP@CT project.

Of Earth, For Earth.